Foreword

Since *Creation Moments* began in October of 1987, I have researched and written well over a thousand programs. Since those first broadcasts, millions of people have heard *Creation Moments* in many languages. Many thousands of listeners have written to thank us for the strength they received through our messages. Our greatest joy here at Bible-Science has come from those who have written to tell us that their "chance" hearing of *Creation Moments* actually caused them to re-think their beliefs about evolution and God. The Holy Spirit has used these challenges to bring some to saving faith in Christ!

Over the years, many of our programs have explored many of the intricate relationships within and between living things. For example, if you were to combine all the information in all the programs we've done about plant communication and self-defense, the result would be a lengthy article — perhaps even a book chapter. Within that chapter, one would learn of dozens of inter-connected details about the very narrow, specialized subject of plant interactions with the rest of the living world. A similar amount of detail about the biochemistry of plants would result in several books.

Details! The creation is filled with more detail than we will ever be able to catalog. As *Creation Moments* has explored some of this detail over the years, my awe over God's ability to design and construct the creation has been immeasurably multiplied.

When Genesis tells us that God said, "Let the earth sprout vegetation . . . ." we are reading about more than the miracle of the creation of plant life. The detailed planning built into that life united more individual designs and purposes than we humans could ever imagine, no less command! God's creating commands in Genesis are clearly much more than just commands.

Large human projects may require years of planning before they are constructed. Did God spend time in eternity, planning out everything before bringing the creation into material reality? After all, the grass of the field is not possible unless the matter from which it is made has electrons with a very specific charge. God had to design the electron to precise specifications or the grasses would not live. As the basis of life, the carbon molecule had to be given special properties or even the most precise electron would not make life possible.

Did God agonize through these countless details before He began the creation? I suspect not. The details that we call planning — no matter how intricate — are part of God's nature and being. As one writer put it, efficiency is one of God's attributes. This means that God created every detail of the universe with the complete picture of the finished creation in mind because such efficiency is simply part of His nature.

I pray that your faith will be strengthened as you read this volume of **Letting God Create Your Day**. And if the idea of God's efficiency is new to you, I encourage you to look for evidences of it in the pages that follow. With that small training, you will also be able to see God's efficiency in the world around you — and in your own life.

Paul A. Bartz
Host, ***Creation Moments***

Letting God Create Your Day

Volume 4

Paul A. Bartz
Bible-Science Association

The Scripture quotations in this publication are from **The Holy Bible, New King James Version,** copyright © 1982 by Thomas Nelson, Inc.

Letting God Create Your Day: Volume 4
by Paul A. Bartz

Copyright © 1993 Bible-Science Association, Inc.
P.O. Box 33220
Minneapolis, Minnesota 55433

Cover: The Bird of Paradise flower is one of the most striking and unusual in the world. Some species are tree-like and reach heights of up to 25 feet. Potted Bird of Paradise plants remain smaller than they would in the wild because their roots are confined, which encourages early flowering. *(Photo courtesy of Paul A. Bartz)*

Library of Congress Catalog Card Number 92-75059
ISBN 1-882510-05-4

All rights reserved. No portion of this book may be reproduced in any form without the written permission of the Bible-Science Association.

Printed in the United States of America.
Printing and production costs for this book were underwritten by Mr. & Mrs. Harold E. Anderson of St. Cloud, Minnesota.

The Wasp that Gives Up its Life

John 15:13
"'Greater love has no one than this, than to lay down one's life for his friends.'"

Evolution says that the relationships between living things developed because such relationships turned out to be beneficial for everyone involved. This is, of course, very different from the principle we find in the Bible of serving and giving ourselves to others. Now, with God as creator, we should expect to find examples of such relationships which are so clear that no amount of imagination can interpret them as other than selfless giving.

One such case can be seen in the relationship between two wasps, *Ibalia* and the *Sirex*. Normally, the *Sirex* grub burrows deep into the wood of a tree, where it changes into an adult who then burrows back out of the tree. But the *Ibalia* grub sometimes becomes a parasite on the *Sirex* grub, gaining nourishment for its own growth from the *Sirex* without killing it. When the *Ibalia* grub, whose jaws are too weak to burrow out of the wood, is ready to leave the tree, the *Sirex* grub does something it never otherwise does — it burrows back out of the tree, taking the *Ibalia* grub with it. As a result, the *Ibalia* grub escapes, and the *Sirex* grub's life is threatened if not ended.

Evolutionists admit that they have no explanation for this! But it is no mystery when we understand the standards of selflessness set by the Creator Himself when He sent His only Son to win salvation from death for us.

Prayer: Lord God, I thank You for Your selfless love for me in sending Your Son. Help me to learn selflessness in my daily life. In Jesus' Name. Amen.

Worship the Creator

Psalm 148:4-5

"Praise Him, you heavens of heavens, and you waters above the heavens! Let them praise the name of the LORD, for He commanded and they were created."

Expressions of worship and praise for God because He is our Creator are found frequently in the Old and New Testaments, but rarly in our modern churches. It was, however, a common part of Christian worship for nearly two thousand years. But over the last century and a quarter — since Darwin's ideas have gained popularity — such worship and praise has gradually become de-emphasized, even in Bible-oriented churches. When was the last time you heard a good sermon on Genesis Chapter One?

It was to remedy this situation that a broad-based group of creationist organizations first issued the call to Christian churches to celebrate Creation Sunday. Today, this group which calls Christians everywhere to praise God for His work of creation, and learn more about that work, includes all of the leading creationist organizations in the world, as well as many churches.

The Bible-Science Association is among those creation groups which is offering special resources — many at no charge — for your congregation's observance of Creation Sunday . Creationists encourage churches to celebrate Creation Sunday on the first Sunday in February. Let us join together in praising of the Creator of all the heavens and Earth and everything in them!

Prayer: Dear Heavenly Father and Creator, help us to be able to thank and praise you for your work of creation and to be strengthened by Your Word to resist all challenges against Your Truth. In Jesus' Name. Amen.

The Snake that Fishes in the Desert

Psalm 104:27

"These all wait for You, that You may give them their food in due season."

The intelligence of the interrelationships between three unrelated desert creatures gives witness to the fact that there is a single, all-wise Creator of all things. And the way in which the side-winder uses these relationships, could hardly have been provided by natural selection.

While the side-winder's weird, whip-like method of movement across the desert sands propels him faster than you might think, he cannot hope to catch one of his favorite foods, the gecko. Sometimes the side-winder can sneak up on the lizard, but as soon as the gecko spots him, it runs off at top speed. Now the side-winder may began plan "B" and collect a little "bait" for desert fishing. The snake buries his entire body in the soft sand, except for the very tip of his tail, which sticks out of the sand, looking like a lone blade of grass. Soon ants discover this potential source of food in the barren desert and begin to explore it. Now it will only be a matter of time before the gecko finds the ants — one of his favorite foods. While the unsuspecting gecko prepares for dinner, the side-winder strikes from beneath the sand, eating the gecko.

Could the side-winder have perfected this neat little trick by any natural device? No. This a witness to the intelligence of our Creator, as well as to His care of all His creatures — even the side-winder.

Prayer: Dear heavenly Father, we see that You care for all You have made. Help each of us to thank You as we see that your care for us goes beyond giving us food. You also provide for our salvation through Jesus Christ. Amen.

Should the Sun Spin Faster?

Psalm 74:16

"The day is Yours, the night also is Yours; You have prepared the light and the sun."

According to evolution, some explanation for the sun, stars, and planets must be found which does not include God as Creator.

Today, evolutionists generally believe that the sun and the other planets of the solar system are related to each other. One of the most popular theories is that the sun and the planets of the solar system each formed as a single cloud of dust and gases were drawn together by gravity. As this material was drawn together, it began spinning faster and faste, like a giant twirling skater pulling his arms toward his body. According to this theory, the amount of angular momentum — which we will simply call "spin" — among the planets must equal the amount of "spin" the sun has.

As we have gradually learned more about the solar system, we have been able to more accurately compute this energy. Figuring in the mass of the planets, their orbits, and the mass and rotation of the sun, scientists have concluded that 98 percent of the solar system's "spin" is in the planets — only two percent, not 50, as expected, is in the sun. So, modern science has confirmed — probably as much as it can in this case — that the Earth was created independently from the sun. Of course we know that Scripture clearly states that the Earth was created four days before the sun. Here a scientific mystery for evolution is no mystery at all for those of us who turn to the Bible!

Prayer: Dear Heavenly Father, Creator of all stars, we give You thanks that You have shown us some of Your great power in creating the stars. Help us to realize that You come to us not to bring fear, but Your undeserved love through the salvation that is found in Your Son. In His Name. Amen.

Anti-Evolutionary Behavior

Psalm 14:1a
"The fool has said in his heart, 'There is no God.'"

When the Bible talks about things in the material world which are part of our experience, we find that the Bible is always faithfully true. But there are many things in the creation which contradict the theory of evolution.

For example, Darwin said that natural selection could never produce anything for the sole benefit of another species. However, Darwin's rule doesn't hold true. Previous *Creation Moments* have presented some examples of creatures that help other creatures to their own detriment.

Today we go to the ants for yet another example. A number of ant species actually help another species to its own harm. Robber ants will take the larvae of a certain beetle into their nests to nurture and raise. Not only do robber ants neglect their own larvae to do this, they will actually feed their *own* larvae to the beetle larvae. For some reason which scientists don't understand, the queen ant in such a colony often becomes sterile, and eventually the entire colony dies.

Evolutionary theory declares that relationships of this kind could not develop. But this week alone we have mentioned two of many exceptions. We wonder whether our all-knowing Creator has not created these relationships for the purpose of confounding men's wisdom when they chose to ignore His work!

Prayer: Dear heavenly Father, we ask that You would not only prevent each of us from ever knowingly or unknowingly denying You, but that You would help us tell others about You and Your forgiveness for us in Jesus Christ. In His Name. Amen.

The 1988 Search for Noah's Ark

Matthew 24:37

"'But as the days of Noah were, so also will the coming of the Son of Man be.'"

Each year Christians eagerly await the latest word from the late summer expeditions looking for the possible remains of Noah's Ark. In 1988 one team was allowed to survey a part of Mt. Ararat by air but found nothing. Another team, however, was the only team allowed to actually work on the mountain itself.

The Snow Tiger team, organized by Dr. Charles Willis, used radar to probe beneath the mountain's ice at one site, and eliminated one suspected location from the list of possible Ark sites. But it was another discovery by the team that deserves the headlines. With the help of their mountain friends, the team discovered, high on the mountain, the remains of what may be the oldest city yet! The remaining huge blocks of stone bear witness to ancient walls, houses and streets. Evidence suggests that this may be the city which Ham, son of Noah, built after the great Flood. The Willis group plans further study.

But the Christian faith does not depend on finding Noah's Ark. Our most powerful evidence is the empty tomb of Christ and the results in our lives. The Ark reminds us that we need to be ready for Christ's return to judge all the Earth — which could happen at any moment. And what is your preparation? Know and believe that Jesus Christ has removed your sin by His suffering and death, and that He gives you a new life to live through His Resurrection.

Prayer: Dear Lord Jesus Christ, I know that You have carried my sins on the cross. Help me to always be ready for Your return and motivate me to help others be prepared. Amen.

Oldest Bug is Modern

Genesis 1:24

"*Then God said, 'Let the earth bring forth the living creature according to its kind: cattle and creeping thing and beast of the earth, each according to its kind'; and it was so.*"

Did you know that the phrase or thought, "according to its kind" appears ten times in the first chapter of Genesis? Not only is the fact that plants and animals reproduce according to their kind an important principle in the Bible, it is central to modern agriculture and husbandry. What's more there is no evidence in the fossil record that any creature ever existed that was in the process of changing from one kind of creature into another. While many different kinds of plants and animals have become extinct, modern forms are found in the most ancient rocks.

This fact was highlighted for us again when scientists announced the discovery of a fossil bug which they said was 15 million years older than any insect ever found before. Of course, we know that those are highly inflated years. But the discovery of the insect will lead evolutionists to revise their story of how insects evolved. And they are not just going to have to allow that bugs have been around for more of the Earth's history than they thought. The scientists also noted that the insect they found was remarkably like modern-day silverfish. In other words, insects were well-established long before evolutionists ever thought they had evolved.

And so, another plank is added to the platform of those who believe, based on Scripture, that so-called modern creatures have been on Earth ever since the first week of its history.

Prayer: Dear Heavenly Father, I thank You that You have given us Your sure Word, especially that I might know the truth of Jesus Christ which leads to salvation. In His Name. Amen.

A Dangerous Wonder

Matthew 10:16
"'Behold, I send you out as sheep in the midst of wolves. Therefore be wise as serpents and harmless as doves.'"

The rattlesnake, though dangerous, is one of the greater wonders of God's creation. The eyes of the rattlesnake are different from all other back-boned animals. Rattlesnakes have no eyelids, but their eyes are protected by clear cover-plates. Their yellow lenses slide in and out, like binoculars — and like built-in binoculars provide detailed magnification of distant objects.

The rattlesnake has many more ways of sensing its environment than we do. Its lower jaw is linked, through fine bones, to its inner ear. With this arrangement, if the rattlesnake places its lower jaw on the ground, it can hear distant footsteps — its jaw acting just like a stethoscope.

Rattlesnakes also adjust their venom dosage to the size of their prey. The larger the victim, the more venom needed — and the rattlesnake knows just how much to deliver. The truth is, about 20 percent of all rattlesnake bites inflicted on humans deliver no venom at all because we are too big for the snake to swallow. While the rattlesnake is dangerous, more people die from bee stings each year than rattlesnake bites.

The abilities of the rattlesnake certainly glorify God, Who designed this amazing creature. Realize that you, too, are fearfully and wonderfully made by the same Creator, and that He has given you even more wonderful abilities to use in grateful service to Him.

Prayer: Dear heavenly Father, I thank You that You have given me abilities that I can use to glorify You. Help me to better understand how I can glorify You more with my life. In Jesus' Name. Amen.

Is it a Boy or a Girl?

Genesis 1:28
"Then God blessed them, and God said to them, 'Be fruitful and multiply; fill the earth and subdue it; have dominion over the fish of the sea, over the birds of the air, and over every living thing that moves on the earth.'"

Researchers have learned that some fish have an absolutely amazing ability — a specialty which is of considerable help to fish and to man as well. This remarkable gift is the capacity to change sex *after* they are hatched!

Scientists have learned that silverside eggs placed in relatively warm water will eventually produce mainly male fish, no matter how many females originally hatch. At the same time, eggs allowed to hatch in relatively cold water will produce mainly females.

This is an advantage because eggs hatching in warm water are usually hatching closer to winter, and young strong males are more likely to survive the winter than young females. On the other hand, eggs hatching in the spring usually hatch in cold water, producing females which have plenty of time to grow before facing winter. Researchers point out that these findings could be of tremendous help to fish hatcheries who wish to manage their populations and production.

Our Creator has wisely provided for the preservation of His creatures. He has designed some of these gifts in such a way that human beings can use them to put the Earth and its plants and animals into service for their, as well as our good. The witness to our Creator could not be more clear.

Prayer: Dear heavenly Father, I thank You that You provide for the needs of all Your creatures. Help me to be a good steward of Your creation In Jesus' Name. Amen.

Diamonds in the Sky?

Matthew 6:19, 21

"'Do not lay up for yourselves treasures on earth, where moth and rust destroy and where thieves break in and steal . . . for where your treasure is, there your heart will be also."

Imagine — countless tons of diamonds, virtually free for the taking. Such wealth is the stuff of legends. But now it might be for real. The only catch is that these diamonds, if they exist, are under an atmospheric pressure as high as six million times that on Earth, and the temperature could be a more than balmy 12,000 degrees (F). Those are the conditions thought to exist on the planets Uranus and Neptune. Scientists now believe, contrary to their earlier speculation, that simple molecules could not exist in normal forms on these planets. This put an end to speculation about life evolving on these planets. Great pressures, along with extreme heat, force all the carbon atoms to be compressed into diamonds.

Why would God create all those diamonds in a place where they don't do anyone any good? The fact is, God knows that material wealth gives us very little. We human beings begin to realize this fact when we discover that all the wealth in the world is meaningless if no one loves or cares about us.

Our Creator is the only source of true wealth. Through His Son Jesus Christ, He brings the riches of heaven to the poorest among us. He brings to His children peace that cannot be destroyed by the hottest war zone, and gives life that never ends. You can learn more about these riches, which bring both temporal and eternal benefits, in the pages of the Bible.

Prayer: Dear Father in heaven, teach me to place the correct value — Your value — on all things and then seek true value through Your instruction on Holy Scripture. In Jesus' Name. Amen.

The Fragile Rattlesnake

John 3:14

"'And as Moses lifted up the serpent in the wilderness, even so must the Son of Man be lifted up....'"

The rattlesnake is a dangerous creature, but surprisingly vulnerable. Believe it or not, rodents, rather than being easy victims of the rattlesnake, are one of the rattlesnake's greatest enemies. Ground squirrels will sometimes attack rattlers, lunging and biting them. Ground squirrels can inflict wounds which become infected and cause the snake's death. They often get away with this because their reflexes are so much faster than the snake's.

Those who handle rattlesnakes must be careful of their bite. But they must be equally careful about how they pick the snake up, because its neck is easily broken. If a rattler is trapped for only 20 minutes in the noon-day sun, it can go into convulsions. And if it is placed on ground with too steep an incline, its heart, unable to pump uphill, may fail. In fact, simply handling a rattler may cause it to starve itself to death.

Scripture refers to the devil as the "serpent" many times. However, just as the rattlesnake is truly deadly, but also very vulnerable, so too is the devil truly deadly, but vulnerable. When Jesus Christ gave His life on the cross in payment for our sins, He beat the devil. All who believe in Jesus Christ as their Lord and Savior from sin, death and the devil have that victory through Christ, even though the devil would like us to forget that Christ has been beaten.

Prayer: Dear Lord Jesus Christ, I thank You that through Your innocent suffering and death You have delivered me from sin, death, and the devil. While I know that in this life I will still have to contend with the devil, always keep me mindful of how You have beaten him for me. Amen.

The Bible's Age for the Earth

Genesis 1:1

"In the beginning God created the heavens and the earth."

Just *when* was the "beginning" when God first created the heavens and the Earth, as spoken of in Genesis 1:1? In these days when scientists talk about "billions" of years, the question of what the Bible says on the matter of origins becomes even more important.

Few people doubt that the Bible intends to teach that the creation is young. While the people who lived in Old Testament times didn't use the same kind of calendars we do, modern research has shown that they did use calendars which can, with some accuracy, be translated into years "before Christ." The genealogies of the Old Testament are actual reproductions of the calendar system used in most ancient times.

While the subject is very complicated, we can share with you the calculations done by some well-known Christians. You are probably familiar with Bishop Ussher's calculation that creation took place in 4004 B.C. But did you know that no less than the great scientist Kepler calculated that creation took place in 3877? Martin Luther calculated that creation took place in 3961 B.C. The very oldest ages arrived at through calculations based on Scripture say the creation is about 7,500 years old.

For the Christian, the question of the age of the creation can only be answered on the basis of Scripture. There is no question that Scripture gives us more than enough information to conclude that the creation is young and that God wants us to know it. After that, it is simply a matter of whether we accept Scripture's authority.

Prayer: Dear Heavenly Father, help my thinking not to be conformed to this world, but help me to be transformed by Your renewing of my mind so that my faith in Your clear Word is unshakable. In Jesus' Name. Amen.

Mt. St. Helens Surprises Scientists

Genesis 2:4
"This is the history of the heavens and the earth when they were created, in the day that the LORD God made the earth and the heavens...."

If the Bible is the history that it claims to be, the features of the Earth, as we know them today, were formed much more rapidly than most people think. As a result, scientists are surprised at the speed at which life is returning to Mt. St. Helens.

In 1980 Mt. St. Helens exploded with incredible force, leaving a sterile, barren scar on the face of the Earth. Some said that it would be generations before life would return to the area. Yet only five years after the eruption, U.S. Forest Service ecologists reported that virtually every species that had lived in the area before the eruption had already returned. In most, but not all cases, the populations were much smaller than before the eruption — but they were back.

Unexpected forces helped cause this transformation. For example, as gophers tunneled, they mixed soil into the top layer of volcanic ash. Seedling trees, protected during the blast by thick snow and the debris of larger trees, provided the basis for a nearly instant, if not tall, new forest which has grown over the few years since the blast. Spirit Lake was supporting fish only five years after the blast.

The return of life to Mt. St. Helens shows us that commonly accepted ideas about how long life takes to establish itself need to be revised downward drastically. And this fact helps to show us that the Bible's claims of a young Earth are not at all unbelievable.

Prayer: Dear Father, I thank You for giving us accurate history in Scripture which we can trust so that we can be made wise unto salvation which is in Jesus Christ. In His name. Amen.

Eating Machine

Isaiah 65:24, 25b
"'It shall come to pass that before they call, I will answer; and while they are still speaking, I will hear. . . . and dust shall be the serpent's food. They shall not hurt nor destroy in all My holy mountain,' says the LORD."

The rattlesnake is such an efficient eating machine that no one has ever gotten a complete look at a rattlesnake's feeding without the aid of special high-speed equipment.

Those famous fangs of the rattlesnake are mounted on hinges so that they can fold up along the roof of the snake's mouth when not in use. Each fang is a hollow needle, with a canal on the side of the end of the fang through which venom is injected. After only a couple of uses, the fang drops out, to be replaced by another.

When it strikes, the snake's neck muscles propel the head with such speed that a car with the same acceleration would go from 0 to 60 miles per hour in just half a second! As the head moves toward the prey, the fangs unfold and a suitable amount of venom is released into the victim. In less than a second the snake is back in position. The venom not only kills the victim, but it also begins tenderizing the meat. When he swallows his victim, the rattlesnake can unhinge its jaws to make the job easier.

Scripture tells us that such violence was not part of God's original plan for the world. It is one of the consequences of sin. But when Christ returns to judge the whole creation, He shall once again provide a perfect world — never touched by sin or evil — for all those who have forgiveness of sin through Him.

Are you ready for His return?

Prayer: Dear Lord, help me prepare for Your return so that I may meet You with joy because I know that I am cleansed of my sin through Your sufferings and death on the cross. Amen.

The Magnificent Cheetah

Job 38:39
"Can you hunt the prey for the lion, or satisfy the appetite of the young lions . . . ."

Of 41 species of cats around the world, the cheetah is one of the most unusual. Unlike other great cats, the cheetah cannot roar, but it can purr like a house cat or emit high-pitched chirps. Breathtakingly beautiful in form, the cheetah is the world's fastest land animal, reaching a speed of 40 miles per hour from a standstill within two seconds, and can momentarily reach speeds of 70 miles per hour. The cheetah's claws are more like a dog's than a cat's because it is unable to retract them.

The Lord has truly given the cheetah a high-performance body. He has an unusually powerful heart, an oversized liver, extra large and strong arteries, and extra large nostrils for taking in great quantities of air. The cheetah also has hip and shoulder girdles which swivel on its spine. As it runs, the cheetah's spine curves up and down as his legs bunch and then extend. When moving at high speed, the cheetah may only touch the ground once every 23 feet.

While most scientists believe that the cheetah evolved, the very oldest cheetah fossils show us an animal that is just about like the cheetahs we know today. This complete lack of evidence for evolution, plus the intelligent specialized features of the cheetah, lead us to the conclusion that the cheetah is a special creation of God.

Surely God must be filled with joy as He watches these magnificent creatures.

Prayer: Dear Father, we thank You for the great beauty and variety You have placed in the creation. As Your Word says, Your power and wisdom are certainly evident around us. In Jesus' Name. Amen.

From the Creator's Hand

Matthew 2:18
"'A voice was heard in Ramah, lamentation, weeping, and great mourning, Rachel weeping for her children, refusing to be comforted, because they were no more.'"

It is difficult to find creationists who are not also pro-life. This is because creationists tend to be more aware than most of how God is intimately involved in our world. There are countless examples of how God has taken special care of each of His millions of species of creatures.

For years we have been talking about these wonders on *Creation Moments*. And as Jesus said, when we see the great care God gives to seemingly unimportant creatures, we should realize how much He cares about each and every one of us.

There are no "extra" people in His world. There are no "unwanted" children, only sinful people who don't want what God is giving them. God makes each and every one of us with purpose in mind, but He never forces us to accept His purpose. Like a loving suitor, He seeks to win us to Himself through the sacrifice of His Son which cleanses us from sin.

While the world looks at statistics, God looks at individuals, lives, abilities, and hopes. We must learn to do so, too. The unborn life, your life, my life — they are all of concern to Him. Not one life is not worth living. But the real meaning of life can only be found in God's Word which tells us about what He has done for each of us through His Son, Jesus Christ.

Put away the bleak wisdom of the world and look to His light in Scripture.

Prayer: Dear Father, help me to see more clearly how You love me. Help me to see more clearly Your individual care in my life. In Your grace and mercy, remove the scourge of abortion from our midst and replace it with love. In Jesus' Name. Amen.

Hunting with the Cheetah

Joel 1:6

"For a nation has come up against My land, strong, and without number; his teeth are the teeth of a lion, and he has the fangs of a fierce lion."

Great speed is the special gift that God has given to the cheetah so that it can make its living. No other cat has such speed, and as a result, the cheetah's tactics are unique, especially since the cheetah will not eat anything it has not freshly killed.

Cheetahs hunt alone, during the day, on the grassy plains. This means that its prey usually sees it coming. The hungry cheetah carefully stalks its prey, hoping not to be noticed until it is 600 to 900 feet away from the victim. At that point the cheetah breaks into a flat-out chase, reaching speeds of up to 70 miles per hour. By the time the cheetah catches up to his prey, he seldom has enough strength left to kill anything very large.

About the largest animal the cheetah will hunt is the 40-pound Thomson's gazelle. And after killing an animal this size, the cheetah will lie panting for half an hour before he is able to eat. During this time, any lion or hyena which happens along may steal the hapless cheetah's hard-earned meal. In one study in Tanzania's Serengeti National Park, researchers found that lions hijacked 12 percent of the cheetah's meals.

The incredible variety of special abilities the Creator has given to different creatures not only shows us how great His imagination is, but also that He cares for every creature He has created.

Prayer: Dear Heavenly Father, You have personally given each creature special gifts and personal love. I pray that You would help me to more clearly understand and better use the gifts You have given me. In Jesus' Name. Amen.

Seeing Heat, Tasting Smells

1 Corinthians 13:12
"For now we see in a mirror, dimly, but then face to face. Now I know in part, but then I shall know just as I also am known."

There is much in the creation that we cannot fully appreciate because we lack the senses which could tell us about these things.

You may have noticed that the rattlesnake has little pits in its face, under its eyes. Amazingly, these pits are a second pair of eyes which actually see what eyes cannot see! Researchers have learned that there is a membrane stretched over the back of each pit, a thousandth of an inch thick, which is crammed with sensors which pick up the infrared radiation, or heat, given off by all warm-blooded creatures. These sensors are hooked up to a second vision center in the brain so that the rattlesnake can see an image of its warm-blooded victim even in complete darkness.

You may have heard that when a snake flicks its tongue out, it is not insulting you. Rather, the snake is smelling you. Its tongue picks up scent molecules in the air, then rubs the molecules off on a sense organ in the roof of its mouth. This provides it with a scent of whatever the molecules came from. You may be amazed to learn that human beings are equipped with the same organ before birth!

There is so much of reality to sense, and our ability to sense it all is very small. This is why it is nothing but human pride to say that if we have not seen something for ourselves, it cannot exist. But the faith that God gives allows us to see His workings which are usually otherwise invisible to us.

Let God's Word sharpen Your senses.

Prayer: Dear Father in heaven, help me to have more clear sight, so that I can see, by the faith You provide, those things that I would not otherwise know about. In Jesus' Name. Amen.

The Secret of Bee Bread

Numbers 13:27
"Then they told him, and said: 'We went to the land where you sent us. It truly flows with milk and honey . . . .'"

Bee bread is a highly nutritional derivative, made from pollen by bees. Scientists have understood for some time that worker bees who have just emerged from the comb must eat bee bread so that their glands produce food for the queen and developing larvae. However, older worker bees get by on honey alone. So scientists suspected that bee bread must have a higher nutritional value than honey.

Now, after painstaking work, researchers have learned how bee bread is made. Even as bees collect the pollen, they begin to work on the recipe. They add secretions from special glands, as well as microorganisms that produce enzymes which release a number of important nutrients from the pollen. Other microbes are added to produce antibiotics and fatty acids to prevent spoilage. At the same time, unwanted microbes are removed, and the bees add honey or nectar to the bread to get it to stick together.

The bees' recipe for bee bread involves highly sophisticated knowledge of microbiology, nutritional chemistry, and biochemistry in general. Modern science therefore absolutely requires us to reject the idea that bees and their culture evolved by blind chance and ignorant natural forces.

Only a perfect Creator could have made these wonderful creatures and taught them how to make bee bread!

Prayer: Dear Father, I thank You for the gift of honey, and for creating clear indications of Your work as Creator that cannot be denied. In Jesus' Name. Amen.

Nobel Laureate Suggests Spirit!

Genesis 2:7
"And the LORD God formed man of the dust of the ground, and breathed into his nostrils the breath of life; and man became a living being."

Many scientists have rejected the idea that man has a spirit — that we are more than simply material organisms. After all, they reason, man evolved, but a non-material thing like a spirit cannot evolve.

Evolution is completely materialistic in its outlook. The first evangelists for evolution admitted that they devised the idea of naturalistic evolution to get rid of the idea of spiritual reality.

That's why it is noteworthy that, in 1983, Nobel prize-winning scientist George Wald announced to the scientific community that he had found evidence of a non-material aspect to life. Wald called this non-material part of life consciousness, and concluded that our consciousness exists outside of space and time.

Evolutionary scientists quickly treated Wald as a heretic. Suddenly evolutionists attacked one of their own with the same ferocity they had previously reserved for creationists. It is interesting to note that none of the material I have seen has actually argued with Dr. Wald's science — they simply refused to even look at his evidence because they just *knew* that he had the wrong conclusion.

The lesson we learn from this is that evolution is a faith — a faith that competes with Christianity. Another lesson is that there *is* real scientific evidence that supports the Christian view of life and reality.

Prayer: Dear Father in heaven, I thank You that You have revealed the truth about who and what we are in Scripture. Help me not to be intimidated by those who promote a competing faith as science. In Jesus Name. Amen.

The Friendly Cheetah

Daniel 6:27
"'He delivers and rescues, and He works signs and wonders in heaven and on earth, Who has delivered Daniel from the power of the lions.'"

From a human standpoint, Daniel would have been a whole lot safer if he had been thrown into a den of cheetahs rather than a den of lions. But because he had the Lord as his protector, it didn't matter what the animals were.

The truth is, cheetahs are the least aggressive of the big cats and offer virtually no threat to man at all. In their native Africa, bushmen actually chase cheetahs away from their kills and steal the fruit of the cheetah's hard-won labor. Man has been taming cheetahs and training them to leashes for thousands of years.

Recently, a naturalist returned a hand-raised cheetah to the African wilds. The naturalist had to train the cheetah to hunt, just as its mother would have done. In the process, the cheetah repeatedly chased antelopes to its human friend, and waited to see what it was supposed to do with them.

The cheetah is a unique variation on the cat theme which shows how God's creativity can take many forms. When we study His creation, we are wise if we remember that, compared to God, we have hardly any imagination. Therefore we could never decide what could be and what could not be.

As we learn about the world God has created, we do best when we study what God has done, without letting our ideas of what is possible and impossible cloud our thinking.

Prayer: Dear Father in heaven, help me to have a faith in Your working like Daniel had, a faith which is not limited by my own thinking of what You can do to help me. In Jesus' Name. Amen.

Christ: The Answer to Superstition

Galatians 3:1
"O foolish Galatians! Who has bewitched you that you should not obey the truth, before whose eyes Jesus Christ was clearly portrayed among you as crucified?"

Christians are often portrayed as foolish and superstitious. Various writers and commentators talk about Christianity as an outdated, prescientific faith. Creationism has often been called *pseudo* or false science. But is it true that Christians are more given to superstition and pseudoscience than others in society?

A study which carefully analyzed figures gathered by the U.S. Census Bureau shows that accusations that conservative, Bible-believing Christians are more prone toward superstition are completely wrong. The truth is just the opposite. The study shows that Bible-believing Christians are, in the report's words, "most virtuous according to scientific standards when we examine cults and pseudoscience proliferating in our culture today." The study also found that the irreligious and religious liberals are most prone to cults and pseudoscience.

Now stop and think for a moment. So-called fundamentalist or Bible-believing Christians almost always accept the Genesis account of creation — yet they are the least prone toward false science. We clearly see the principle in history. When biblical Christianity takes over among people, they have less interest in superstitions and cults. And creation is very much a part of that Christianity.

Yes, Christ and the gospel is the light of God in this dark world!

Prayer: Dear heavenly Father, I thank You that the light of the gospel of Christ has shone in my life because of Your Word. Help me to be able to spread that light to others in this world which is rushing toward so many superstitions and cults. In Jesus' Name. Amen.

Hot Flowers

Song of Solomon 2:11-12

"For lo, the winter is past, the rain is over and gone. The flowers appear on the earth; the time of singing has come, and the voice of the turtledove is heard in our land."

At the end of winter, when so many of us become weary of winter and look forward to spring, it might seem a little early to be talking about flowers blooming. You might like to know, however, that man is not the only one of God's creatures who would like to welcome spring a little early. There are some 2,500 species of plants that actually generate heat when they flower, and many of them are heating up beneath northern snows.

As early as February, in western Pennsylvania, the skunk cabbage begins to generate enough heat to melt the ground and snow around itself so that it can begin to grow and flower. Even if the air temperature is 0 degrees (F), the skunk cabbage can raise the air temperature around its flowers to 50 degrees — the temperature it needs for flowering. Plants in the same family begin heating up across the northern climates around the world.

Sometimes it seems as though the entire living world strains toward spring at this time of year. In the midst of widespread fears that the weather has somehow gotten seriously off track, spring should remind us of what the Lord promised Noah and his descendants — all of us, and all living things — "While the Earth remains, seedtime and harvest, cold and heat, winter and summer, and day and night shall not cease."

Prayer: Dear Father in heaven, I thank You that You comfort me and all living things with the promise that You will preserve the conditions necessary for life on Earth until the day that Your Son Jesus Christ returns to take all the faithful home with Him. In His Name. Amen.

Mercury's Young Magnetic Field

Genesis 1:14-15

"Then God said, 'Let there be lights in the firmament of the heavens to divide the day from the night; and let them be for signs and seasons, and for days and years; and let them be for lights in the firmament of the heavens to give light on the earth'; and it was so."

In Psalm 19:1 we read, "The heavens declare the glory of God." When we read this we should be reminded of more than the witness of the beauty of the skies. The material creation bears witness to God's glory in many different ways.

Take, for example, the claim made by modern evolutionists that the creation is billions of years old. This claim started with those who consider the Bible to be an out-dated collection of ancient myths. Everyone realizes that a young creation has to be the work of a Creator. Yet there are many ways in which the creation makes its own witness to its youth, and therefore to the glory of God.

In 1974 and 1975, the space probe Mariner 10 studied the magnetic field of the small planet Mercury. Scientists expected that since the solar system is supposed to be 4.5 billion years old, Mercury would have lost all its magnetic field long ago. Yet, scientists discovered that Mercury's magnetic field was alive and well! Today, after nearly fifteen years, scientists have been unable to offer an explanation. But the explanation is very simple.

If Mercury is only about 6,000 years old, which is consistent with the Bible's history of things, we can easily expect that if Mercury was created with a magnetic field, it should still have some magnetic field left. And so, tiny Mercury, which doesn't seem to have much use otherwise, clearly declares the glory of our Creator God!

Prayer: Dear heavenly Father, help me not only to glorify You with my lips, but also through all my actions, so that what I am invites people to love You. In Jesus' Name. Amen.

Mystery of the Elephant Ears

Psalm 103:15

"As for man, his days are like grass; as a flower of the field, so he flourishes."

There are a wide variety of decorative and ornamental plants that go by the popular name "elephant ears" because of their very large leaves. Elephant ears are among the 2,500 known species of plants that generate heat in much the same way as warm-blooded animals. Some flowering elephant ears actually raise their temperatures close to human body temperatures while in bloom!

Even more amazing is the relationship between many elephant ears and the scarab beetle. It has only recently been learned that many elephant ears depend completely on the scarab beetle for pollination. Elephant ears produce a flower stalk, called a spadix, which has three different kinds of flowers, rnale, female and sterile. The beetles are drawn to the ready flowers by the sterile flowers, which they love to eat. In the process of crawling around on the spadix to get to these flowers, the beetles also pick up pollen from the male flowers and pollinate the female flowers. Yet, none of the male or female flowers are ever eaten, even though they look exactly like the sterile flowers!

This close dependency between totally unrelated creatures which have special features perfectly suited to one another bears witness to an overall plan and Planner for all the creation. And if our Creator has made such detailed provisions for these plants, which are here today and soon gone, you can rest assured that He has made many more provisions for you and all of mankind.

Prayer: Dear Father in heaven, help me to see, by faith, how You have cared for all the details of my life, beginning with Your plan of salvation for me through Jesus Christ. In His Name. Amen.

What Mothers Always Knew

Psalm 8:2
"Out of the mouth of babes and infants You have ordained strength, because of Your enemies, that You may silence the enemy and the avenger."

You wouldn't guess that a 36-hour-old infant would be able to teach grown scientists about the Creator — especially by making faces. But that is exactly what has happened! The Bible teaches that each of us is an individual, made by God, and known to Him even before we are born. However, those who have tried to promote the idea that chance and natural law can explain without the Creator, what the Bible explains *with* the Creator, have portrayed the newborn as basically a blank slate. They contend that newborns are completely without most of the basic attributes which we consider human. To these people there is no such thing as normal human expressions like a smile —everything is learned.

But now researchers, studying 36-hour-old babies, have learned differently. They now confirm that smiling is a natural, human reaction — something which babies do even *before* they are born. Babies also recognize human faces and can tell the difference between human expressions. Babies also connect expressions they see with their own expressions. Even the newborn in a crib is not kicking and crying without meaning. What's more, scientists have confirmed one other thing that mothers have always known — babies really enjoy those silly faces and funny sounds we use to get them to laugh.

The Bible clearly offers a better view than modern humanistic evolution when it defines each human being, from before birth, as a unique individual, made by a Creator Who loves us and seeks our love.

Prayer: Dear heavenly Father, I thank You that You have made me and that You assure me that You love me and seek my love. Help me to believe Your promises to me, and to show my love for You in everything I do. In Jesus' Name. Amen.

Bird Brained

Ecclesiastes 10:20

"Do not curse the king, even in your thought; do not curse the rich, even in your bedroom; for a bird of the air may carry your voice, and a bird in flight may tell the matter."

All of us are familiar with birds that talk. Mankind has been teaching birds to talk for thousands of years. Man has also been using birds to deliver messages for just as long. But new research on the intelligence of birds suggests that they might know more about the messages they carry than we ever suspected.

Researchers studying pigeons have learned that they can read all of the letters of the alphabet and they read them in a way similar to people. Scientists became convinced that pigeons see the world in much the same way as we do when they discovered that pigeons tend to confuse the very same letters that people do. What's more, it took only four months for pigeons to learn to distinguish all the letters of the alphabet. For years pigeons have been used to spot and reject defective items on assembly lines, where they have better records than human beings.

Based on evolutionary theory, scientists had always believed that recognition of abstract forms like letters requires large brains like those found in humans and so-called higher primates. But common experience and Scripture have viewed animals as more intelligent than evolutionary science has because evolutionists, rejecting an intelligent Creator, sees evolution as the source of intelligence.

Real science continues to disprove evolutionary views.

Prayer: Dear wise heavenly Father, I thank You that You have so generously provided intelligence to the animals, not only that they can make their livings, but also so they can serve man help us. In Jesus' Name. Amen.

Shifty Red-Shifts

Deuteronomy 4:19
"'And take heed, lest you lift your eyes to heaven, and when you see the sun, the moon, and the stars, all the host of heaven, you feel driven to worship them and serve them, which the LORD your God has given to all the peoples under the whole heaven as a heritage.'"

All of us remember hearing about how many of the stars and other objects in space are millions and even billions of light-years away. Since the light from those objects is slightly redder than expected, scientists say that they are moving away from the Earth at tremendous speeds. So the redder the light, the more distant they are. But is the red-shift really proof that there are objects millions of light years away?

Creation scientists point out that many things can cause light to become red. Various dust and gases like those found in space regularly turn our sunsets and sunrises red. Astronomer Halton Arp of the Mt. Wilson and Las Campanas Observatories in Pasadena, California has uncovered many evidences that the red-shifting of starlight is more likely caused by factors other than distance. I should add that Dr. Arp is not a creationist, just an honest scientist. And his work has caused a great stir among scientists who believe in evolution.

When man uses evidence from the creation — like starlight — to try to prove that there is no God, he is placing the creation above the Creator. In other words he is practicing idolatry. We Christians must be careful before accepting any claims not found in the Bible. And we must reject those claims which contradict the Bible.

Prayer: Dear heavenly Father, I thank You that You are greater than all things in the creation. Help me to better learn Your Word so that I am never trapped into the idolatry of any "learning" from the creation that goes against Your Word. In Jesus Name. Amen.

Wrasse Confusion

Job 12:8, 13
". . . or speak to the earth, and it will teach you; and the fish of the sea will explain to you. . . .With Him are wisdom and strength, He has counsel and understanding."

God's unlimited imagination and exuberant creativity really had the biologists puzzled for a long time. Had they not rejected the Creator, when they saw the flamboyant colors of the family of fish called the wrasse, and began running into mysteries, they might have guessed that the Creator put a little extra imagination into these creatures.

In the early days of marine biology, scientists often identified male and female wrasse of the same species as totally different species. They don't even look like they are related. Biologists noticed that they had a mystery on their hands when they discovered that they had ended up designating some species that had only males, and others which had only females. Eventually, observation cleared up the confusion — only to result in more confusion! Scientists found certain wrasse who looked neither like males nor females. Scientists also found that schools around some small reefs had no young males. It seemed as if the males just appeared, fully grown, without ever growing up. And in a way, they had. For among these wrasse, when there is a shortage of males, one of the females will turn into a fully functional male!

In Scripture, the word "understanding" often refers to a deep and detailed knowledge of how different things work together. The special and highly creative arrangements that God has provided for the wrasse are an excellent example of God's understanding as well as the unlimited range of His imagination!

Prayer: Dear Father in heaven, I thank and praise You for Your great beauty and inspiring imagination that is so evident all around us in the creation. Through the instruction of Your Word, grant me the gift of understanding. In Jesus' Name. Amen.

Biological Clowning

Genesis 1:21-22
"So God created great sea creatures and every living thing that moves, with which the waters abounded, according to their kind.... And God saw that it was good. And God blessed them, saying, 'Be fruitful and multiply, and fill the waters in the seas....'"

When God made fish and told them to be fruitful and multiply, He gave some of them very special abilities so that they could carry out His will. Some 23 families of fish actually change from females to males, or from males to females, to help their species reproduce.

This ability to change sex, as do tropicals like the clownfish, produces some amusing behavior. Clownfish are particularly popular among fish hobbyists, who might find that the female fish they recently purchased has become another male! Wrasse often spawn on large coral reefs, dominated by stronger, territorial males who chase weaker males away from the reef. But often the smaller males will take on the appearance of a female. Under this disguise, they join the line of females that are waiting to spawn with the dominant male. While in line, these pretenders encourage the ladies in waiting to spawn with them, rather than with the large dominant male.

This biological clowning poses a couple of problems for believers in evolution. For one thing, it encourages the very opposite of reproduction of the strongest and best. And there are so many different kinds of fish which change sex that evolutionists have decided that they must have evolved several times.

Of course, we know where the ability to change sex came from — from an imaginative Creator Who decided to follow the same amazing theme with many different kinds of fish.

Prayer: Dear Father, I look in wonder and awe at the beauty and imagination You expressed so freely in Your work of creation. Help me to never cease glorifying You as my gracious and wonderful God! Amen.

Mistletoe's Special Needs Supplied

Genesis 1:11

"Then God said, 'Let the earth bring forth grass, the herb that yields seed, and the fruit tree that yields fruit according to its kind, whose seed is in itself, on the earth'; and it was so."

When God designed the mistletoe, He brought the design of the plant, its way of life, and the feeding habits of birds together into a wondrously designed working system. Then He made one amazing exception to the design.

While the wild mistletoe has chlorophyll so that it is able to make some of its own food, it is a parasite which also receives much of its nourishment from the tree on which it grows. The seeds of the mistletoe must sprout into the living tissue of a tree.

Mistletoe blooms in early winter, producing a white berry. These berries contain seeds as well as a very sticky pulp which birds find quite tasty. Many of the seeds are thus spread through the birds' droppings. Some seeds, covered in the sticky pulp, end up on the outside of the birds' beaks. We have all seen how birds will clean their beaks by rubbing them on a tree branch. If they have been eating mistletoe berries, their beak cleaning glues the mistletoe seeds to a perfect spot for them to grow.

The dwarf mistletoe is the one exception to this system. When its seeds are ripe, the berries explode, shooting seeds to neighboring branches up to 40 feet away.

Whether we look at how God has brought different creatures together for their mutual benefit, or see the creative means He has designed so they can reproduce, we must exclaim, "Truly the Lord has done all things well!"

Prayer: Dear Father, every part of the creation shows forth Your glory. I pray that You would open people's minds so that they may see Your glorious works and join Your people in praising You now and forever. In Jesus' Name. Amen.

To Keep the Station Open

Psalm 51:2
"Wash me thoroughly from my iniquity, and cleanse me from my sin."

Just because fish live in water doesn't mean they stay clean. Parasites and other unhealthy scale and skin conditions can develop unless fish are kept clean.

One of the most unusual arrangements for keeping fish clean is found on the Great Barrier Reef of Australia. That's where fish known as the cleaner wrasse offer a cleaning service for other fish. Each cleaning station is made up of one male and a harem of females. Other fish come and wait in line for their turn to be cleaned.

The only fish not allowed at the cleaning station are male cleaner wrasse from other cleaning stations. The male jealously protects his harem. But when he dies, something most unusual happens. Within an hour of his death, the largest female in the harem begins to behave like a male. Within a couple of weeks, she has become a fully functional male, and head of the cleaning station. Without this change, the cleaning station would have to shut down, leaving a lot of unsatisfied customers to the perils of disease. Here, the Lord has provided an astonishing arrangement for the health of His creatures.

Even more astonishing was the solution which God provided to cleanse us of our sins and make us spiritually healthy. When He sent His Son Jesus Christ to suffer our punishment for our sin, He was providing the only means possible by which we can be washed of our sins and given a healthy new life in fellowship with Him.

Prayer: Merciful Father, I thank You that You have sent Your Son Jesus Christ, Who sacrificed Himself on the cross for my sake. Help me always to depend completely on what He has done for me, rather than on my own good efforts and intentions. Amen.

Termite Assassin

Genesis 1:25
"And God made the beast of the earth according to its kind, cattle according to its kind, and everything that creeps on the earth according to its kind. And God saw that it was good."

There are a few kinds of insects which disguise themselves in order to avoid predators, and some birds and primates do use food as lures. But the assassin bug uses a combination of camouflage and baiting which is totally unknown in any other creature.

The assassin bug prepares to make his living by gluing bits of termite nest to himself using fluid from special glands in his body. This termite nest material is his camouflage as he performs his work at the nest. By the time he is done, no part of his body will be showing.

Sneaking, unnoticed, to an opening in the nest, the assassin bug waits for some hapless termite to come by. Having captured his first victim and drained it of body fluids, the assassin bug now takes his victim and waves it near the nest opening. Seeing the corpse, a worker moves toward the door to retrieve it. As he does so the assassin bug slowly draws his bait back, luring the worker further from the protection of the nest. Suddenly the assassin bug grabs his new victim and discards the old one. Victim number two will become bait for victim number three. In one such operation, researchers watched as one assassin bug continued this for three hours, killing 31 termites!

The gruesome way in which the assassin bug makes his living demonstrates a great deal of intelligence. The assassin bug is yet another example of the fact that intelligence doesn't have anything to do with evolution — rather, it is a gift of the Creator to His creatures.

Prayer: Dear Father in heaven, I thank You for the gift of intelligence. Help me to use the intelligence You have given me wisely, and to Your glory. In Jesus' Name. Amen.

Insects in the Snow

Psalm 147:16-18
"He gives snow like wool; He scatters the frost like ashes; He casts out His hail like morsels; Who can stand before His cold? He sends out His word and melts them; He causes His wind to blow, and the waters flow."

People who live in climates where there is ice and snow are used to seeing the activity of warm-blooded creatures even in winter. However, reptiles and amphibians, because they are cold-blooded, seem to disappear in the winter. But did you know that many kinds of insects continue to be active in the winter?

How can insects be active this time of year without freezing? After all, when living tissue freezes, expanding ice crystals destroy tissue membranes. Insects survive freezing temperatures in much the same way as does an automobile — with antifreeze! Many insects manufacture either a form of alcohol or glycerol, which is chemically similar to the antifreeze in your car.

Eastern tent caterpillars are very much alive within their cocoons, and by midwinter their body weight is over one third antifreeze. One parasitic wasp larvae is able to continue living at temperatures as low as -52 degrees F. Some 25 of the 700 known species of pygmy locust actually live in or near Arctic regions. And the snowfly gets its name from the fact that its life is carried on in the snows of winter.

It seems doubtful that the world before the Flood had winters as we do today. Yet, in His wisdom, God built special provisions into many of His creatures so that they could survive when winter cold as we know it today became a reality.

Prayer: Dear Father in heaven, I thank You that You desires to care for all of Your creatures. Help me to depend more on You and less on myself, so that Your glory may be seen in my life. In Jesus' Name. Amen.

Spiders, Too, Hurt

Proverbs 30:28
"the spider skillfully grasps with its hands, and it is in kings' palaces."

Evolution has given people today the idea that there are "higher" and "lower" animals. As a result, many people have come to think that the so-called "lower" animals don't, for example, feel pain. Some people even fault the Bible for using language which speaks of animals experiencing life as humans do.

But as our knowledge of the created world grows, even the Bible's language on this subject is being vindicated. When a spider catches an insect in its web that is usually the end of the insect. But researchers were watching as a poisonous ambush bug, caught in a web, was about to be bitten and wrapped by a spider. As the spider moved in on its prey, the ambush bug stung one of the spider's legs. To the researchers' surprise, the spider reacted immediately by dropping its injured leg and running back to the center of its web.

This led scientists to wonder if the spider reacted like this because he felt pain. Further study showed researchers that spiders do indeed feel pain, and *most* living things probably do, too.

Since living things are capable of experiencing life in similar ways, even our emotions and senses can be compared to God's — even though they are poor when compared to His. For this reason we should not be surprised to learn that His other creatures experience life as we do without the false limitations imposed by evolutionary distinctions of so-called "higher" and "lower" life.

Prayer: Dear loving heavenly Father, I thank You that You have loved me with a perfect love even though I cannot return perfect love. Help me to be more like You in all my thinking and feelings. In Jesus' Name. Amen.

Older Farmers

Genesis 2:15

"Then the LORD God took the man and put him in the garden of Eden to tend and keep it."

Despite the fact that there are ants who garden and harvest, standard evolutionary theory says that man only developed farming recently in his history. Besides going against common sense, this evolutionary idea is beginning to cause problems with the greatly inflated evolutionary dating system.

A few years ago, evolutionary scientists believed that man was not farming or harvesting more than 8,000 years ago. However, while studying a site in Egypt which, by inflated evolutionary dating was thought to be 18,000 years old, evolutionists found grains of wheat and barley. They couldn't believe that man was farming so early in his history, so evolutionary scientists did more research. Using different methods, they arrived at dates anywhere from 4,800 to 18,000 years old. Not knowing what to do, they decided that the grains must have somehow fallen to the older layer, and that man really wasn't farming or harvesting so early in history.

However, this explanation didn't solve all of their problems. Evolutionists still have no explanation for the hundreds of grinding stones — used to grind grain — which were found at the *older* site. There are two lessons in this for us. One is that evolutionists' dates are not very objective. They pick and chose whatever fits their theory.

The second lesson is that man was farming and harvesting a lot earlier than most people think. In fact, the Bible tells us that the job of the first man, Adam, was to tend the Garden of Eden!

Prayer: Dear Father in heaven, I thank You for the sure and certain revelation of Your Word and that its wisdom and truth confound all who doubt it. Help me to read it more, and better study and understand it. In Jesus' Name. Amen.

Sherlock Holmes' Greatest Mystery?

James 3:16
"For where envy and self-seeking exist, confusion and every evil thing will be there."

In 1912 Charles Dawson announced the discovery of Piltdown man to the world. Piltdown man was a half-human, half ape-like creature which was offered as proof that man was nothing more than a glorified ape. Today everyone knows that Piltdown man was a fraud. Someone had stained an orangutan's jawbone, filed it to fit a modern human skull, and planted the jaw and a skull in the English gravel pit where Dawson was working. But unanswered to this day is the question, "Who dun it?"

Some believe that the hoax was the work of none other than Sir Arthur Conan Doyle — the creator of Sherlock Holmes. Doyle was well-known for his practical jokes. He only lived seven or eight miles from the site where the bones were discovered. and he had visited the site. Doyle was also a doctor who understood human anatomy, chemistry, and anthropology. In addition, he had access to bones like those which were found. What's more, in Doyle's **The Lost World**, his character actually talks about faking bones, and Doyle's map of the "Lost World" looks surprisingly like the Piltdown site and surrounding area.

William Fix, who is not a creationist, argues that no one has yet found one bone to support the idea of human evolution. But, he says, there has been no shortage of those who hype their findings and try to make a name for themselves. Truly, when we place ourselves above God by saying that we are smart enough to contradict the Word of God, the result is going to be confusion and evil — the very opposite of the gifts given by God's Word.

Prayer: Dear heavenly Father, please do not allow the darkness and confusion of empty human strivings and pride obscure Your truth in my life. In Jesus' Name. Amen.

Ever Curious Man

Genesis 1:14
"Then God said, 'Let there be lights in the firmament of the heavens to divide the day from the night; and let them be for signs and seasons, and for days and years . . . ."

The Bible pictures mankind as intelligent, able to learn about the creation and teach others from the first day of his existence. According to Genesis, the first men were even building cities, inventing musical instruments, crafting in bronze and iron, and keeping calendars. Calendars are among the very earliest forms of writing found by archaeologists today.

The Bible's picture of advanced, intelligent and ever-curious man is further supported by the discovery, all over the world, of ancient observatories. Some of these observatories, such as stonehenge, are so ancient that little is known about the builders. In fact, when many of the structures which are now known to be observatories were first discovered, no one could figure out the purpose of these structures. Recently 60 additional observatories were positively identified in the American Southwest. The first, and most famous, is the so-called "sun dagger" in New Mexico. Two spiral designs had been cut into the rock so that a shaft of light, passing through nearby rocks, would illuminate key portions of the carvings only on solstices and equinoxes. These 60 additional sun markers have been identified, some 1,300 years old.

The creation itself declares the glory of the Creator Who reveals Himself more fully in the pages of the Bible. But man's own activities also provide a strong witness to the fact that the Bible's story of the special creation of intelligent, inquisitive man, is completely reliable.

Prayer: Dear heavenly Father, I thank You that You have made me and given me the ability to learn about the world You have made. Help me to always use my mind and abilities to glorify You. In Jesus' Name. Amen.

Don't Offend Your House Plants

Psalm 1:1, 3

"Blessed is the man Who walks not in the counsel of the ungodly . . . . He shall be like a tree planted by the rivers of water . . . and whatever he does shall prosper."

A few years ago there was a lot of publicity about people who claimed that their house plants did better when they were talked to. Many experts thought that this was a silly idea since plants are very low, comparatively speaking, on the (imaginary) evolutionary scale. Research eventually proved that plants which were treated to music, especially classical music, did indeed seem healthier.

New research has convinced many scientists that plants also have feelings and memory. Researchers carried out their experiments on the bur marigold, which typically sprouts and immediately produces two identical leaves or cotyledons. Scientists made several tiny needle holes in only one of the cotyledons, and then, after only five minutes, they removed both cotyledons. Later they removed the top growth of the young plants in order to force growth from the buds adjacent to the cotyledons. As a result, the plants tended to show growth only on the side where the undamaged cotyledon had been.

Further study has shown that, while plants do not have a nervous system, they do have a chemical communication system which works something like memory!

We certainly must marvel at our Creator's care in giving plants feelings and memory. And then we are reminded that if He has taken such care in creating and continuing to care for plants, how much more must He care for each and every one of us!

Prayer: Dear Father, I give You thanks that You have made me in such a way that I can hear and believe in Your grace given through Your Son, Jesus Christ. Help me to see and know Your personal care and love for me every day of my life. Amen.

Man and Horse Domestication

Genesis 47:17a

"So they brought their livestock to Joseph, and Joseph gave them bread in exchange for the horses . . . ."

Bones can tell stories. An interesting story of man's relationship with the horse is told by the fossilized teeth of ancient horses. You see, domesticated horses have a habit called crib biting — biting the rails or even a nearby rock in their corral. Wild horses don't do this. Crib biting creates a specific pattern of wear. So, when scientists find horses' teeth which have this pattern, they know that it comes from a domesticated horse.

At one time it was believed that horses were first domesticated — using inflated evolutionary years — about 8,000 years ago. But several instances of horse teeth with the crib biting wear pattern have now turned up in much more ancient sites. Man may have had a partnership with the horse more than ten times longer than was previously thought. But we who believe the Bible are not surprised that the evolutionary description of man's relationship with the horse must be revised.

In Genesis 47:17 we read that during the famine in which Joseph was administering Egypt's grain, he exchanged grain for livestock, including horses. Bible critics have long said that the horse was unknown in Egypt at Joseph's time — more proof, they said, that the Bible is full of mistakes. But now we know that it was the critics of the Bible who were wrong — man's relationship with the horse is much, much older than most people ever thought, just as the Bible says.

Prayer: My dear heavenly Father, I thank You that Your Word is faithful and trustworthy. Build and strengthen my faith through Your Word and help me to be better able to tell others Your wonderful news of grace and forgiveness in Jesus Christ. Amen.

Another All-Wet Theory of Evolution

Genesis 1:27

"So God created man in His own image; in the image of God He created him; male and female He created them."

There is a universe of difference between saying that man was created by God, in the image of God, and saying that man is nothing more than a hairless ape. And the fact that man is not covered with hair like the ape has led to a weird new twist in the theory of evolution.

British author Elaine Morgan has written that current evolutionary theory, which says that man's ancestors lost their hair because they ran and hunted in the hot sun, doesn't make any sense. If that were true, she says, lions wouldn't have any hair either. She thinks ape-like creatures lost their hair for the same reason that evolutionists believe whales lost their hair — they became aquatic creatures. She also says that swimming apes would have to learn to speak instead of communicating with sign language because their hands would be occupied with swimming. Morgan suggests that with the absence of fossil evidence for missing links between apes and man, her theory makes as much sense as any other.

We would agree with her about the lack of fossil evidence, and that her theory makes as much sense as any other evolutionary theory — and that's no sense at all! Man was made by God to be so different from the ape that it's impossible to say that the two are related. And without fossil evidence, it's not even scientific to say that man is related to apes. Rather than seek a relationship to apes, we are to seek our relationship with God the Father, through His Son, Jesus Christ.

Prayer: Dear heavenly Father, Help me in all aspects of my life to look to You and Your Word for my understanding and faith. Show me where I err in looking to the creation for my hope instead of to You. In Jesus' Name. Amen.

The Venus Flytrap

Psalm 104:24a
"O LORD, how manifold are Your works! In wisdom You have made them all."

The exotic Venus flytrap is a wonder of God's engineering with ingenious alternatives to the way plants usually operate. For example, these meat-eating plants usually live in mineral-poor soils, but by catching their own lunch, they actually provide their own fertilizer.

Up until now, scientists have not fully understood what causes the trap of the flytrap to close so rapidly on its victim. They knew that when an insect trips the little trigger hairs inside the trap, those hairs send an electric impulse to the cells on the outside of the trap.

Now it has been learned that this impulse almost instantly causes the outer cells of the trap to secrete acid. The acid breaks down the cell walls and they expand at high speed, causing the trap to close. The more the insect fights the trap, the more tightly it closes. Six to twelve hours later, after lunch is digested, the trap receives chemical signals from inside — the carnivorous plant's version of a burp — and opens to await the next meal.

The Venus flytrap is a glowing example of what Scripture is talking about when it says that the works of the Lord are manifold — that He has made so much variety, and so many ways of providing for living things. The Venus flytrap could have been created to provide for its needs as do other plants. Truly, the Lord's imagination and creativity are a wonder to behold!

Prayer: Dear Father in heaven, I praise and thank You for how Your work so fills me with joy and wonder. I especially thank You for Your wonderful plan of salvation which has redeemed me from sin, death and the devil. In Jesus' Name. Amen.

When is an Answer Not an Answer?

Acts 17:5

"'Nor is He worshiped with men's hands, as though He needed anything, since He gives to all life, breath, and all things."

As modern science has learned more about molecular biology, it has also learned that life could not have arisen on Earth from non-life. The fact that life could not have come about by known natural laws is so clear that even many evolutionists now admit this truth. However, many of them are now looking to space for an answer to their problem.

Dr. Francis Crick, co-discoverer of the structure of the DNA molecule, calls the origin of life "almost a miracle." He is one of those evolutionists who believes that although life could not have developed on Earth, maybe it could have developed somewhere else. His theory, called "directed panspermia," says that life could have been seeded on Earth by beings who developed elsewhere in space. But this is not really an answer. It simply pushes the question of the origin of life to "somewhere out there in space." How did *that* life start? The important point is that so many evolutionists are now admitting that life could not have started here on Earth without the action of some intelligent agent, unknown to science.

This is exactly what the Bible teaches, for it clearly says that life was created through the instrument of the Word of God. The source of life, and all creation, is found in that creating Word, made flesh for our salvation. This is a much higher answer than saying that little green men from outer space made us. Read the details for yourself in the first chapter of John's gospel.

Prayer: Dear Father, I thank You that You have not only given me biological life, but that You have prepared life eternal for me through Jesus Christ, Who is the creating Word made flesh for my salvation. In His Name. Amen.

The Dormant Toad

Isaiah 43:20

"The beast of the field will honor Me, the jackals and the ostriches, because I give waters in the wilderness and rivers in the desert, to give drink to My people, My chosen."

Few deserts are more inhospitable than the Sonoran Desert of North America. Life is so hard there that even the Couch's Spadefoot toad, despite its remarkable features, must lie dormant for 11 out of 12 months.

Normally the Spadefoot has only one month out of twelve to carry out the normal business of life. The Spadefoot has an internal clock which tells it when the violent desert rainstorms are near and when to begin to near the surface of the ground. The toads, extremely sensitive to ground vibrations, can hear the pounding rains miles away. The distant rains bring the toads out of the sand so that as soon as pools have formed, the male toads are in them, calling for the females. Mating and egg-laying are completed by morning, and dawn finds the toads safely protected from the sun's heat beneath the sand. But the desert pools reach 100 degrees and will not last long before they dry up, killing the eggs. Under ideal conditions it only takes nine days for the eggs to hatch into survivable young toads, leaving anywhere from days to weeks — in a good year — for the toads to eat the food needed to live 11 months until the next rain.

There is no known force in the universe which can create life, even under the most favorable conditions. Our Creator God not only creates life which can survive under the most extreme circumstances, but He is the source of eternal life, as Holy Scripture teaches us.

Prayer: Dear Father in heaven, even though life itself can, at times, seem like an empty desert, help me to remember that Your care for me is much greater than for any of Your other creatures, and that only You are the Source of true life. In Jesus' Name. Amen.

Glacier Races

Psalm 148:4-5

"Praise Him, you heavens of heavens, and you waters above the heavens! Let them praise the name of the Lord, for He commanded and they were created."

It only seems right that life depends on what chemists call one of the most complex substances in the universe. This stuff refuses to act like other materials — and because it does, we are alive. The design of this strange stuff is a tribute to the power of the Creator to design materials which behave as He wishes.

This odd material flows when it is solid, and its solid state is less dense than its liquid state. It can become a gas directly from a solid. Based on the expected behavior of matter, it should boil and freeze at much higher temperatures than it does. In addition, this substance conducts electricity in a very unusual manner — the usual method would not support life.

This mystery substance is ordinary water. If it didn't float when it became a solid, most of the life in every lake in a cold climate would be wiped out every winter. Water's unusual method for conducting electricity makes the processes of life possible within your cells. Water's unique surface tension causes those pretty beads on a newly polished car. And water's strange behavior as ice under high pressure causes glaciers to flow down hill. While many glaciers move at a rate of only one or two feet per day, one glacier was once clocked at 250 feet per day!

Every substance that God has created was made with special abilities and features to make life possible. This is yet another testimony that our Creator God's attentions are focused on *us*.

Prayer: Dear Father in heaven, I thank You that all of the creation has been made by You with us in mind. Help me never to take Your love and care for me for granted. In Jesus' Name. Amen.

King Catfish and Ancient Writing

Exodus 17:14

"Then the LORD said to Moses, 'Write this for a memorial in the book and recount it in the hearing of Joshua . . . .'"

In the history books he is known as Narmer, the first pharaoh of Egypt. His name simply means "Catfish." From what we can tell, King Catfish, who lived around five hundred years after the Great Flood, is helping to prove the truth of the Bible's history.

It was long thought that at the time of King Catfish, or even at the time of Moses, 400 years later, people were unable to write. But recent discoveries of writing on both carved stone slabs and papyrus-like paper from the time of King Catfish have shown that writing had been around for a long time, even at this early date. Even more amazing has been the discovery, from Catfish's time, of a well-organized industrial complex which had many production facilities, each specializing in the production of certain types and quantities of fired clay pots. The ruins clearly show that this complex shipped its product all over the ancient Egyptian empire.

There is even evidence that the Flood itself was recent history to these people. At the time they lived, the still-draining flood waters made Egypt a verdant land of forests and grassy plains. That these people gave their animals a section in the city's graveyard could be related to their recent memory that animals were of enough value to be saved aboard the Ark.

One thing is clear. Those who said that the Bible was wrong when it said that Moses was able to write have, themselves, been proven wrong by old King Catfish!

Prayer: Dear heavenly Father, in this ever-changing world, where so many think that truth changes, I thank You that You have given me Your sure and true Word in Holy Scripture. In Jesus' Name. Amen.

The Riddle of Kemp's Ridley

Psalm 119:140-141

"Your word is very pure; therefore Your servant loves it. I am small and despised, yet I do not forget Your precepts."

God counts things very differently than man counts things. Even if being small and unnoticed usually means unimportance to men, to God being small never means being unimportant. One good example is the smallest of the sea turtles, the Kemp's Ridley, which weighs only 60 to 100 pounds.

So little has been known about Kemp's Ridley that it was not until 1947 that scientists knew that mature members of the species were found only on a beach on Mexico's coast. Scientists are now attempting to establish a second colony in another location. Because the eggs are prized by humans as well as coastal creatures, the population of the Kemp's Ridley is down to less than a thousand egg-laying females. It doesn't help that the Kemp's Ridley is the only sea turtle to lay its eggs only in the daytime.

However, the Creator has given Kemp's Ridley some special advantages to offset these disadvantages. The female can lay fertile eggs for up to several years after she has mated. The hole she digs for her nest is 18 inches deep, and she may lay as many as 135 eggs at one time.

Although Kemp's Ridley was almost completely ignored and unknown to science, God has not treated Kemp's Ridley as unimportant, nor has He neglected to give this creature special gifts for its survival. And though we Christians may be considered unimportant by the decision-makers of this world, God does not treat even one Christian as unimportant.

Prayer: Dear loving heavenly Father, I thank You that You do not value things in the same way as does sinful man and that therefore no one of us, no matter how small and unimportant in the world's eyes, may be neglected by Your love. In Jesus' Name. Amen.

Rat Language: More than Meets the Ear

Psalm 19:1, 3
"The heavens declare the glory of God; and the firmament shows His handiwork. . . . There is no speech nor language where their voice is not heard."

It was once thought by many that language was a relatively late development, even for a sophisticated creature like man. These people said that when the Bible spoke of the whole creation praising God, the Bible was quite wrong, or was just speaking figuratively. Now modern science is learning that many creatures have their own language — language which we point out, can also be used to praise the Creator.

Most humans can hear sounds which are anywhere between 20 and 20,000 vibrations per second. Many creatures that we once thought were silent are busy communicating in sounds which are far higher pitched than we can hear. Young lemmings often cry at frequencies as high as 140,000 vibrations per second. The Norway rat communicates at 70,000 vibrations per second. Scientists, using special equipment, have cataloged many of the calls and have concluded that rats have the richest vocabulary of calls with specific meanings. Young rat pups screech a distress call which reaches the same volume as a jack hammer. Yet because of its high frequency, we cannot hear a thing. There is an advantage for these small rodents having such a high-pitched language. Even though many predators, like cats, can easily hear these frequencies, the loudest call does not travel far enough to give away the location of the nest.

Language is a gift from our Creator Who made all things through the Word. Therefore it is only fitting that the entire creation has language with which to praise Him!

Prayer: Dear Father in heaven, help me to use my lips and voice in praising and thanking You at all times. In Jesus' Name. Amen.

Dinosaurs Need Friends, Too

Proverbs 18:24
"A man who has friends must himself be friendly, but there is a friend who sticks closer than a brother."

Almost every creature seeks companionship with others. This universal, basic need, gives us additional insight into the mind of the Creator.

Those who believe in evolution have always thought that only warm-blooded creatures valued companionship and love. So for generations, dinosaurs, which are supposed to be much further down the evolutionary scale of living things, were seen as hard-hearted, dull-witted, uncaring creatures which didn't even care whether their own young lived or died. Now, many instances of the desire for companionship, even among dinosaurs, is being uncovered. Scientists are learning that dinosaurs often lived together in communities, building nests, caring for their young, and traveling together, even when they had no need for protection. In some sites, huge dinosaur trackways show that various kinds of dinosaurs lived in groups, close to each other, and that the same kinds generally followed their own pathways. These discoveries, at several sites and involving many different kinds of dinosaurs, have convinced scientists that dinosaurs were very much like modern animals in their social habits. In other words, dinosaurs needed friends too.

That almost all of His creatures value companionship tells us a little about our Creator. Scripture reveals more details about this aspect of the Creator when it tells us that He created man because He wanted someone to love. Do you know His love for you through Jesus Christ?

Prayer: My loving heavenly Father, I thank You for my friends. Most especially do I thank You for Your love for me in sending Your Son Jesus Christ to be my truest and best friend and brother. In His Name. Amen.

God's Optic Fibers

Job 12:7a, 8a, 9
"'But now ask the beasts, and they will teach you . . or speak to the earth, and it will teach you; and the fish of the sea will explain to you. Who among all these does not know that the hand of the LORD has done this. . . .'"

One of the areas in which modern technology is advancing most rapidly is fiber optics. Crystal clear telephone messages are carried on cables of special fibers which conduct messages in digital form using laser light. Computer scientists are designing computer systems in which processors are linked with fiber optic cables in order to increase speed. But man has only *discovered* this technology — he has not invented it, for God used fiber optics in providing for the needs of certain plants.

Scientists have learned that sprouting oat, corn and mung bean plants use the principles of fiber optics so they can grow. When the tip of a young seedling first pokes its nose above ground, it gathers and sends light to the growth center which may be half-an-inch, an inch, or more below the ground. The conducted light then stimulates the growth center, providing energy for growth. Without this provision, the seedling could die, and surely would be held back. Scientists have learned that the seedling stem is such an efficient optic fiber that it will even faithfully conduct a focused image of the light source.

We are learning that just about every technological breakthrough made by man was already used by the Creator when He made the creation. There is hardly a stronger argument than this in favor of belief in an intelligent Creator!

Prayer: My dear, wise, heavenly Father, Your wisdom is so far above ours that what to us is a great achievement is simpleton's logic to You. Enlighten me with Your gracious gifts, that I may know more of Your wisdom and apply it in my life. In Jesus' Name. Name.

Electric Bushes and Trees

Job 14:7, 9
"'For there is hope for a tree, if it is cut down, that it will sprout again, and that its tender shoots will not cease. . . . yet at the scent of water it will bud and bring forth branches like a plant.'"

Are plants insensible? Is it silly to speak of dry plants anticipating a good watering? Is it incorrect to see plants as anticipating spring? Scientists are learning that although plants do not have a nervous system, they still use electricity to respond to their surroundings — and sometimes respond just like human beings!

Do plants respond to touch? Most of us are familiar with the Venus flytrap, which closes its trap when the trigger hairs are disturbed. Then there is the "sensitive plant," which closes its leaflets in response to being touched. One mimosa tree was so disturbed by its keepers that it shed all its leaves, seemingly, as one writer put it, having a nervous breakdown. Tomato plants will wilt to conserve water, but if they are overly disturbed, they will wilt even though they don't need to save water. Scientists have found that cells within plants communicate with each other through electrical signals in much the same way as do our nervous systems. Several different plant activities have been associated with these signals. Such responses are very unexpected in what evolutionists consider simple plants.

However, if we see the creation as the work of a caring Creator Who endowed all of His creation with its own form of emotional sense, even these limited plant responses are not unreasonable reflections of Him. In the importance He has given emotions, we will hopefully see the importance of His love.

Prayer: Dear heavenly Father, I thank You that You so loved the world that You gave Your only Son so that He might earn my salvation through the forgiveness of sins. in His Name. Amen.

Feathered Pilots

Job 12:7, 9

"'But now ask the beasts, and they will teach you; and the birds of the air, and they will tell you... Who among all these does not know that the hand of the LORD has done this....'"

You are the pilot of a Boeing 767 airliner, bringing the great ship in for a landing. Approaching the airport, you extend the slat on the leading edge of each wing so that you have more lift as your speed slows. As the airport nears, you raise the nose of the plane to catch more air. In the final moments of landing you bring the feathers, which act as air breaks, out of the wing to reduce your lift and settle easily onto the runway. If you are like most people, you would never want to bring a great airliner in for a landing, even if you were offered all the free training you needed.

Yet without any training, young birds do this same thing every day. High speed photography has revealed that birds follow each one of these sequences as they approach and land. First, as they prepare to land, they slow their speed and increase their lift by extending and lowering the front part of the wing. Then, to slow even more, they angle their body from a horizontal flying position to a more upright angle, and finally they drop the feathers on the back of their wing — after which the plane's *feathers* are named — to settle nicely on a tree branch, with virtually no effort at all.

Yes, the principles of flight, refined by over 80 years of man's experience, were simply *taught* to the birds by our loving Creator. What a wonderful testimony birds offer to the work of our wonderful Creator!

> *Prayer: Dear creating and saving Heavenly Father, all man's combined experience and learning throughout history does not begin to compare to Your understanding of all things. Help me to remember this as I hear man brag to himself about all he has accomplished. In Jesus' Name. Amen.*

The Power of God Touches Earth

Psalm 22:1a and Mark 15:43b
"'My God, My God, why have You forsaken Me?'"

Good Friday is the day in which the hand of God was more visibly active on Earth than on any other day between the Great Flood and the Resurrection of Christ. His Hand was so clear that even the Roman centurion could see it. But the incredible Hand of God was even more evident to those who knew the Old Testament Scriptures.

King David, described by Scripture as a man after God's own heart, despite his sinful weaknesses, most clearly saw the Savior crucified for his sins a thousand years before it happened. In Psalm 22 he not only described the details of the Savior's crucifixion, but he also quotes the very words which would come from the Savior's mouth, and the words which those around the cross would say — a thousand years in his future! He describes, in detail, what it felt like to be crucified, even though the torture had not yet been invented. He described Christ's pierced hands and feet, and even how the soldiers divided the Savior's clothes and cast lots for His robe.

Even though this amazing record, accurate in every detail, was widely distributed in writing long before the event, most scientists today — who so value prediction to prove accuracy — see His death for our sins as only a normal death, or even a myth. But God's Hand in human history could not be more clear than it is here. That detailed description of Christ's death for your sin and mine, a thousand years before it happened, is so dramatic because it is God's greatest desire that you believe what He has done for you on the cross of Calvary.

Prayer: Dear loving Father, I thank You that You sent Your Son to pay the grievous penalty of my sin. I thank You that You have called me by Your gospel and made me Your child through the forgiveness of sins. In Jesus Name. Amen.

Reversible Wings: Mystery or Creativity?

Isaiah 31:5
"Like birds flying about, so will the LORD of hosts defend Jerusalem. Defending, He will also deliver it; passing over, He will preserve it."

Sometimes incredible wonders pass right before our eyes but we are not equipped to see them. This is often just as true for wonders in the material realm as it is for spiritual wonders.

Take the hummingbird. It has an incredibly long tongue that wraps all the way around its head in a channel under the skin. The extra length allows it to be stretched extra long to probe deeply into trees for food. This feature is not really evident when you look at the hummingbird. Nor is the wonderful way in which its wings move, so that it can hover, fly forwards or fly backwards, its wings beating so fast that it makes the characteristic humming sound.

High speed photography reveals some amazing secrets about the hummingbird's wings which cannot be noticed by the eye. All birds, except the hummingbird, move their wings at the shoulder, elbow and wrist. Only the hummingbird is capable of turning his wings upside down, so that when he hovers, both strokes of his wings beat the air in order to support him. This demonstrates that the hummingbird is not a modification of the general bird design — as evolutionists tell us. Rather, the hummingbird was created with unique features which enable it to make its living.

Likewise, when God tells us that He cares for those who are His through faith in Christ's saving work, we don't always see exactly how He is caring for us. But like the virtually invisible features of the hummingbird, His care is nevertheless there.

Prayer: Dear Father in heaven, I thank You that You have made me Your own child through Jesus Christ. Help me to believe what You tell me in Your Word, even if I cannot see it all for myself. In my Savior's Name. Amen.

The Monkey Puzzle Tree

Psalm 72:18
"Blessed be the LORD God, the God of Israel, Who only does wondrous things!"

The familiar Norfolk Island Pine is part of a very strange family of plants which offer special examples of God's imaginative creativity.

The Norfolk Island Pine is a member of a family called *monkey puzzle*. They are native to southern South America, New Guinea, New Caledonia and, of course, Norfolk Island, an island in the Southern Pacific Ocean. There are several different species of monkey puzzle trees which show, more clearly than the Norfolk Island Pine, where the name monkey puzzle comes from. The Monkey Puzzle Tree is so called because its pattern of branching is so irrational that you can't tell where limbs begin or end. Another member of the family, found in Australia, is also called the Monkey Puzzle Tree because its needles are so arranged that monkeys are left totally befuddled about how to climb it. One odd feature of the Norfolk Island Pine makes it unwise to try to start a new plant from cuttings. If you take a cutting from the vertical shoot at the top of the tree, you will ruin its appearance If you take a cutting from a side shoot, it will root, but it will always grow horizontally, never forming a new tree.

The early pioneers of science understood that God was not forced to create anything in a certain way — there was no limit on His creativity. So they saw science as an effort to study, first hand, just how God *chose* to do things, or as one great scientist said, "to think God's thoughts after Him."

Prayer: Dear heavenly Father, Your thoughts are so far above ours. Help me to learn about Your thoughts and how You work as I study Scripture so that I can better understand Your thoughts. In Jesus' Name. Amen.

Is it Plant or Animal?

Psalm 93:5

"Your testimonies are very sure; holiness adorns Your house, O LORD, forever."

As we learn more about God's creation, we find that a number of things we learned in school aren't really true, and never were. We learned in school that plants are plants and animals are animals, and there are always distinct differences between them. We also learned that complex things like eyes are found only in higher, more evolved creatures. One-celled creatures are very simple.

Dinoflagellates are considered by biologists to be among the simplest forms of cellular life. Yet these amazing creatures are far from simple. Both zoologists, who study animals, and botanists, who study plants, claim that dinoflagellates belong to their area of study. In other words, these little fellows are so complex that scientists can't even figure out whether they are plants or animals. Many dinoflagellates — even though they are but a single cell — have a sense organ like an eye.

Some dinoflagellates cause the deadly red tide which often kills fish. Others give off a red dye during the daytime which is harmless and by night they produce brilliant displays of phosphorescence. Many have two grooves, each of which is equipped with a whip-like appendage for swimming.

While they are small, even these so-called simplest forms of one-celled life are by no means simple. Man's word, even in matters of science, can never be as sure as God's Word about the fact that He made all things. As God's Word says, every creature was created in finished form.

Prayer: Dear Father in heaven, I thank You that Your Word can be trusted. Help me to see more clearly how man's word, no matter how sincere, can never be as sure and certain as Your Word. In Jesus' Name. Amen.

The Too-Young Rings of Uranus

Revelation 6:14
"Then the sky receded as a scroll when it is rolled up, and every mountain and island was moved out of its place."

You have no doubt seen pictures of the planet Saturn and its beautiful rings. Astronomers have learned that most, if not all, of our solar system's outer planets have rings. But as beautiful as these rings are, they are a problem for those astronomers who think that the universe is billions of years old. You see, these rings, which so beautifully encircle the outer planets, would not be around today if the universe was really billions of years old.

Recent studies on the rings of Uranus highlight the problem. A number of theories have been offered to explain why these rings might still be around. One theory says the rings are kept in shape, and in orbit, by small satellites circling the planet near one set of rings. The problem is, there are nine other rings around the planet which are not associated with satellites.

Astronomers have also found 50 to 100 tenuous dust bands around the planet. These dust bands would disappear even more rapidly than rings. One astronomer has suggested that perhaps these bands are replenished with dust when tiny grains of dust collide with invisible moons around the planet.

Science usually doesn't have room for invisible moons and unseen causes. It is unfortunate that when the only alternative is to admit that the Bible is right about the creation being relatively young, and that it will come to an end in man's lifetime, some scientists are so biased that they must invent invisible moons and mysterious causes.

Prayer: Dear heavenly Father, teach me now through Your Word, and so prepare me for Your Son's return to earth. In His Name. Amen.

Who Invented the Book?

Luke 4:17
"And He was handed the book of the prophet Isaiah."

At the time of Christ, they did not have bound books as we do today. Have you ever wondered who gave us the wonderful invention of the modern book, and why they did it?

When Christ walked this Earth, large written works were made up of pages, each connected to the next, which were finally wound around a wooden or metal post. The whole works was called a scroll. If you wanted to read the first page, you had to unroll only a small amount of this awkward arrangement. But if you wanted to read the last page, you had a lot of unrolling and rolling on your hands. About the only benefit to the scroll was that it was a lot easier to handle than clay tablets. So you can understand how the modern bound book was a major advancement in written communication.

Who made this advance? And why? The modern form of the book came about when Christians decided that if they cut the pages apart and bound them together on one edge, they would have a much easier way of reading and finding things in their Bibles. Recently, scholars identified the oldest bound book ever discovered. It was found buried with a child who died 1,600 years ago, and it was a wood and leather bound copy of the Psalms. So next time you hear someone say that Christians are against education, or that Christians don't appreciate books, just remember that Christians and their love for the Bible are responsible for the convenient form of our modern books.

Prayer: Dear Father, I thank You for the love of Scripture that You have worked in the generations of believers before us, as well as the examples of their love for Your Word. In Jesus' Name. Amen.

A Mammoth Mystery

2 Peter 1:20-21
"Knowing this first, that no prophecy of Scripture is of any private interpretation, for prophecy never came by the will of man, but holy men of God spoke as they were moved by the Holy Spirit."

One of the most important differences between man's word and God's Word is that man's word can sometimes be true, while God's Word is always true. This difference is important to remember when scientists start talking about history they never experienced.

When a scientist finds the fossil imprint of a fish, he usually says that he has found a fish. He will probably even give it a name. But what has he really found? The flattened and deformed imprint of what used to be a fish's dead body. Everything else the scientist says about the imprint is story-telling. The stories might be true, but the important thing to remember is that stories are not facts. The uncertain nature of such story-telling was recently illustrated when scientists decided that the Holly Oak pendant, a whelk shell on which a mammoth is sketched, is not really evidence that man saw mammoths in North America. Some dating systems supported an ancient age for the shell. But now it has been decided, based on other dating systems, that the shell is only 1,000 years old, and that the carving is modern.

What is the Holly Oak pendant? It is a shell with a mammoth carved on it. Anything else we might say about it is simply story-telling. But God's Word is different. Because the history and other matters explained to us in Scripture are God's Word, it is completely trustworthy — and man's story-telling about ancient history can never disprove anything that Scripture says.

Prayer: Dear heavenly Father, I thank You for Your sure and true Word. Help me to better appreciate the wonderful gift of Your Word and to study it more seriously. In Jesus' Name. Amen.

Not Like Father, Like Son

Genesis 1:21a

"So God created great sea creatures and every living thing that moves, with which the water abounded, according to their kind . . . ."

It seems that when it came to designing the millions of different creatures in the sea, God used an additional measure of imagination. Part of the reason might be that creatures which live in the near weightlessness of water can be designed with more variation in the shape of their bodies and in the ways they can move. But some of the weirdest forms cannot be easily explained even in this way.

It almost seems as if the Lord made a few creatures for the sole purpose of providing examples of animals that will not fit into any explanation which denies that He is the Creator. Evolutionists have, for years, used the developing embryonic stages of living things to try to support their theory. Even though new knowledge keeps proving that they are on the wrong track, some very adamant evolutionists continue to use the comparisons. But sometimes embryonic forms are so different from the adult forms that the only explanation can be the unlimited imagination of an all-powerful Creator. The adult form of the *Sacculina* barnacle looks more like a fungus than any barnacle. It was only by studying the embryo of the *Sacculina* that scientists were able to learn that this weird creature is really a barnacle.

So next time you watch a nature program about the sea in which they are busy talking about evolutionary development, keep in mind that there is much they aren't going to tell you. And often what they don't tell you are the embarrassing facts which point to our Creator!

Prayer: Dear heavenly Father, I stand in awe as I marvel at Your unlimited imagination, expressed through the creation. Help this to lead me to trust Your wisdom in my life, too. In Jesus' Name. Amen.

Prayer Puzzles Scientists

Ephesians 6:18
"Praying always with all prayer and supplication in the Spirit, being watchful to the end with all perseverance and supplication for the saints."

Amazed researchers recently reported in the *Southern Medical Journal* that prayers can help sick hospital patients, even if they don't know they are being prayed for. But some scientists not involved with the study said they would not accept the findings unless healing on the order of a miracle took place.

Most, but not all, scientists are evolutionists. As evolutionists, most of them believe that since science cannot prove that the God of the Bible exists, or that man has a spirit, the spiritual aspect of man does not exist. They reject talk about man's relationship with God. Yet a number of studies on Christian prayer have been done and they all seem to show that prayer works. In this latest study, 12 percent of those prayed for were better off than those who were not prayed for.

As Christians we need to be careful in thinking about scientific studies on prayer. We don't pray because science says it works. We pray because our loving heavenly Father has commanded us to pray and because He has promised to hear us. Nor does God treat people like statistics in an experiment — each one of us is an individual. And finally, we shouldn't think of prayer as something that forces God to do anything — what is done is always His will, not ours. But the fact that we can verify that prayer works gives us a point to use in witnessing that God truly desires a relationship with each one of us.

Prayer: Dear Father, I thank You that You hear and answer my prayer as if I were the only one praying, even though millions may be praying at the same time. In Jesus' Name. Amen.

Quick Oil

Genesis 7:21
"And all flesh died that moved on the Earth: birds and cattle and beasts and every creeping thing that creeps on the Earth, and every man."

Does oil really take millions of years to form out of once-living materials? Many people think so because that's what evolutionists have been telling them. But creationists either believe that petroleum was created by God at the formation of the Earth, or that it was formed after the burial of living organisms in the Flood.

Because of the influence of evolution, many people have a hard time believing that petroleum could be formed rather easily in a short period of time. Yet a process for doing just this has been in use for over ten years. Recently Canada's Environment Minister, Tom McMillan, announced a $196 million project to build a commercial scale plant which would turn sewage into oil. The plant would use heat to turn Halifax's half-million tons of sewage per year into 700,000 barrels of oil per year. The entire process, essentially the same as that which formed current oil reserves, takes only 30 minutes! This helps demonstrate that living things buried beneath the rapidly-deposited sediments of the Flood 4,500 years ago, and subjected to heat, could easily have produced today's oil deposits.

The Apostle Peter wrote about our day when he pointed out that people would deny that there was ever a world-wide Flood. They don't want to hear that God judges sin. But we also must remember that God gave His only Son so that, believing in His work for us, we will be saved from the consequences of our sin.

Prayer: Dear heavenly Father, I thank You that You are merciful and forgiving toward me for the sake of Your Son, My Lord Jesus Christ. I pray in His Name. Amen.

Against All Odds

Ephesians 1:7
"In Him we have redemption through His blood, the forgiveness of sins according to the riches of His grace."

Blood is one of the most miraculous of all creations. It carries oxygen and energy to our cells, and carries off wastes. Blood is one of the body's communications pipelines, using powerful hormones to provide communication between various parts of the body. Hidden in your blood is the story of nearly everything, good or bad, that is going on in your body.

Hemoglobin is the basis of blood's wonderful ability to carry oxygen to your cells. Many different kinds of creatures, including lobsters and spiders, all have some type of blood with hemoglobin in it. Some creatures even have transparent blood. Hemoglobin, all by itself, testifies to a Creator. You see, so many different and obviously unrelated creatures have hemoglobin that evolutionists could only account for this by saying that hemoglobin must have evolved many times in many different creatures. This explanation worked when scientists thought that hemoglobin was a relatively simple molecule. But now we know that hemoglobin is a very complex, eight-helix twisted molecule of about a hundred atoms, all arranged in just the right way around a central atom of iron. There is a zero chance of this complex molecule happening accidentally even once, no less the many times suggested by evolutionists.

Blood is a wonderful creation of God. He wants us to realize this so that we look for a relationship with Him, which is possible for us through the shed blood of His Son, Jesus Christ.

Prayer: Dear Father in heaven, I thank You that You draw me to Yourself so that I may be assured of Your forgiveness and acceptance through the sacrifice of Your Son. In His Name. Amen.

Similar Needs, Similar Designs

1 Corinthians 15:39
"All flesh is not the same flesh, but there is one kind of flesh of men, another flesh of beasts, another of fish, and another of birds."

A **Nova** program, aired on public broadcasting stations around the country, included in its presentation the argument that human evolution is proven by the fact that chemically, humans and apes are 99 percent alike. That argument appears often. Dr. Duane Gish of the Institute for Creation Research answers it by pointing out that since a cloud is 99 percent water, and a watermelon 97 percent water, the watermelon must have missed being a cloud by only 2 percent.

Actually, the argument that there is only a 1 percent difference between man and apes assumes that we know a lot more about man and apes than we really do. And some of these comparisons can be very misleading. If you take insulin, your insulin does not come from apes, but from beef cattle. Beef insulin is nearly the same as human insulin. Nor do these comparisons hold up when we compare other creatures that are supposed to be related. Similar comparisons show that the mouse is less like the guinea pig than it is like a chicken, a rattlesnake, or the bonito fish. And the bonito is less like the toadfish than it is like a chicken, duck or even a human being!

The old saying that statistics can be used to prove anything certainly applies to this evolutionary argument. The real reason that similar creatures often have similar chemical structures in their bodies is that creatures with similar bone structures tend to have similar needs and so have been given similar chemistry by the Creator.

Prayer: Dear heavenly Father, while I marvel at Your great wisdom in the creation, fill me with a greater desire to learn Your revealed wisdom in Holy Scripture. In Jesus' Name. Amen.

Does the Bible Describe Fossils?

Job 26:5
"Dead things are formed from under the waters, and the inhabitants thereof."

Does the Bible describe fossils? Before you answer too quickly, you need to know that fossils are not a modern discovery.

Even before the birth of Christ there were thinkers who were trying to explain the origin of the world and all living things through natural processes in an attempt to avoid God. They knew that fossils were produced from once-living creatures. And they suggested that fossils were evidence that evolution had taken place. Fossils are produced when a plant or animal dies, and is covered by water or air-carried material before it has a chance to decay. We would naturally expect that the great Flood would have produced many fossils. Job 26:5 seems to be talking about both fossil formation and the resulting fossils when it speaks of dead things forming under the waters. If Job was written, as many think, within a few centuries after the Flood, it is very possible that the fossils which are so deeply buried now were much easier to find then. The connection is so obvious that even the 1837 edition of **Matthew Henry's Commentary** suggests that Job is here talking about fossils. The commentary even includes a drawing of a plesiosaurus fossil. Many more modern commentaries take an evolutionary view or ignore the subject.

While we cannot be absolutely certain that Job is talking about fossils, we do know that the Bible does have much to say about the material world — and when it does so, it is always accurate!

Prayer: Dear heavenly Father, I thank You that Your Word is always accurate in all that it talks about so that its accuracy in worldly subjects might draw some to its accurate promises about a relationship with You through Jesus Christ. In His Name. Amen.

Wisdom in Hiding

1 Corinthians 1:7
"But we speak the wisdom of God in a mystery, the hidden wisdom which God ordained before the ages for our glory."

You don't have to watch a drama very long, or read too far into a book, before you find the hidden being revealed. God's Word often talks about truths which are hidden because of unbelief. But people forget that one important consequence of the hidden is that the truth often comes as a surprise.

A scientist was recently studying insect-eating birds in an oak tree which was in bloom. He was surrounded by male flowers called catkins. All of a sudden, one of the catkins began walking away. The surprise which he had discovered was a species of caterpillar which feeds on catkins, and ends up looking just like a catkin, thereby escaping the notice of feeding birds.

With further study the scientist learned that the members of this species which hatch in early spring and feed on catkins end up looking like catkins, right down to having fake pollen sacks. Members of the same species which hatch later end up looking like oak twigs, and are equally hidden in the summer branches. Even more surprisingly, it is the caterpillars' diet that causes the difference in appearance.

When things are hidden, reality is always different than we expect. While the Gospel is clear, Scripture often mentions that unbelief causes the clear to be hidden to the unbeliever. It is helpful for us to understand this as we see the popularity of belief in evolution and a world without God.

Prayer: Dear Father, I thank You that Your Word is clear. Remove all unbelief from me so that I may see Your truth more clearly. In Jesus' Name. Amen.

Fast Bees or Fast Talking?

Genesis 1:25
"And God made the beast of the Earth according to its kind, cattle according to its kind, and everything that creeps upon the earth according to its kind. And God saw that it was good."

According to the Bible, God created the land-living insects, which would include bees, on day six of creation. This was just three days after He created the flowering plants. According to evolution bees began their evolution from some other insect after pollen-bearing flowering plants had developed. The specialized bee and its complex social structure was thought to have taken a long time to develop.

Creationists, who believe the time line of events laid out in Scripture and reject inflated evolutionary years, were not surprised when it was announced this past year that a bee was found preserved in amber which was many millions of years older than the oldest known bee. Of course we don't accept the inflated evolutionary years. The almost perfectly preserved bee is like modern bees, and can even be identified as a worker. Not only does this show that bees, with all their superb specializations, have been around much longer than ever thought by evolutionists, it also shows that they were around for some time before this specimen lived. In fact, the evolutionist who reported the findings admitted that there is a real problem explaining how bees could have developed nearly at the same time as pollen-bearing plants.

As creationists, we have no such problem. Bees were created, fully formed, only a few days after pollen-bearing plants, by our Creator. Science has shown us once again that the biblical history makes more sense than the evolutionary story of history.

Prayer: Dear Father, the whole creation shows forth Your glory. Open the eyes of those who are blind to Your fingerprints in the creation so that ultimately they may be led to know You and Your forgiveness in Jesus Christ. In His Name. Amen.

What are Your Trees Saying?

Psalm 96:12
"Let the field be joyful, and all that is in it. Then all the trees of the woods will rejoice before the Lord."

Can trees communicate with each other? When Scripture talks about different things in the creation praising God, is it possible that trees really can praise God? Or is this just figurative language?

We live in an age which is quick to dismiss as figurative anything in Scripture which it does not understand, or which disagrees with its values. This is largely because of the influence of evolutionary naturalism. Unfortunately, sometimes we Christians are a little too ready to do the same. After all, trees cannot communicate, can they? Actually, science has known for some time that trees are able to communicate with each other. It was always thought that trees communicated with each other by releasing chemicals through their root systems. But now a physicist in Oregon has accidentally made a discovery which may explain how trees communicate with each other.

While Ed Wagner was studying the flow of sap in trees a couple of summers ago, he started finding voltage readings in the trees. With more research, he discovered that trees generate a standing wave formation which not only travels in the tree, but also through the air to other trees. These waves, he finds, correlate with what the tree is experiencing. The blow of an axe not only causes a tree to send out strong waves, but the surrounding trees respond with a similar, but less emphatic reaction.

Can tree communication also praise God? Now that we know that trees literally talk to one another, there is every reason to believe that they also literally praise their Creator, just as Scripture says.

Prayer: Dear heavenly Father, I stand in awe and wonder at the wisdom and power of Your creation. Let me never neglect to praise You and thank You for Your goodness. In Jesus' Name. Amen.

The Efficient Fish

Genesis 1:21
"So God created the great sea creatures and every living thing that moves, with which the waters abounded, according to their kind.... And God saw that it was good."

Fish are typical of the elegant designs found in the creation. Both function and form are united with the mechanical and biological needs of the fish to produce intelligent, beautiful, yet simple solutions for their needs.

Fish are efficiently designed so that water, containing oxygen, is taken in through the mouth. This is the very point on a fish where inward pressure is greatest. The used water is expelled from behind the gill flaps, the very point at which outward pressure from the fish's body is the greatest. The result is the most efficient system possible — many fish need expend no extra energy to breath when they swim.

Likewise, fish tend to have their eyes located at the point on their bodies where water pressure while swimming is zero. This is important since the curvature of the cornea of the eye determines the focus of the fish's vision. It could be disastrous to the fish if its vision changed as swimming speeds varied. The fish's heart is located over one of the points in its body where the outward pressure is greatest. While the heart muscle works by contraction, the outward pressure on the surrounding surface of the fish allows easy re-expansion of the heart for the next beat.

If fish had gears and pulleys instead of their much more sophisticated biological machinery, no one would deny it was carefully engineered by an outside intelligence.

Prayer: Dear heavenly Father, I am constantly amazed at Your wisdom which I see at every level of the creation. Help me to remember that You have cared much more for me than any other creature. In Jesus' Name. Amen.

Holding Comparisons at Arm's Length

Psalm 139:13
"For You have formed my inward parts; You have covered me in my mother's womb."

Many people still believe that all of us, in our development from fertilized egg through to birth, go through the evolutionary stages which led to human beings. They believe that at one point in our development we are much like fish. Later, many think, we are more like apes, even having a tail. Unfortunately, this myth — based on scientific fraud which was uncovered in the early 20th century — continues to be taught in some textbooks as fact.

Some modern evolutionists continue to argue that evolution is demonstrated by the fact that our developing arms are constructed on the same pattern as most lower creatures. They say that our arms even go through the evolutionary stages, looking like the fins of a fish at one point. Many people are fooled by these arguments which are not even close to the truth. The fact is, the human embryo's arms are always human arms. Their tissue, and the tissue from which they develop, is genetically nothing other than human. And while they may not look like an adult's arms as they develop, there are also great differences between a baby's arm and a construction-worker's arm. What's more, despite the similarity in engineering between a human arm, a lizard's foreleg, a seal's flipper or a bird's wing, each of these creature's forelimbs develops from a different part of the embryo.

As Christians in a modern, scientific age, we need not ever be afraid of what true science will discover. True discovery clearly bears witness to God! We expect it!

Prayer: Dear Father, truly I was was fearfully and wonderfully made by You in my mother's womb, no less than was Adam formed by Your very Hand. Thank You for making me. In Jesus' Name. Amen.

Can Cats See in the Dark?

Psalm 145:15
"The eyes of all look expectantly to You, and You give them their food in due season."

Vision works in a lot of different ways in different creatures. Some creatures can see forms of light which are invisible to us; others have built-in binoculars. The rattlesnake has two sets of eyes, one set for light that is visible to us, and another set for infrared radiation. But one thing that creatures with vision all have in common is that sight requires some form of light.

One of the most unusual experiences in life is total darkness. While it is very difficult to find a place, even at night, where there is total lack of light, millions of people have experienced complete darkness under the careful guidance of a cave tour guide.

While some creatures, like cats, are supposed to be able to see in the dark, they cannot, in fact, see any better than we can in a room which is too dark. However, cats, and many other animals are able to see more clearly than we can in light levels where our eyes become useless. One of the main reasons for this is that their eyes have a reflective layer of cells below the light receptors. So, if a particle of light — called a photon — misses a light receptor on its way into the eye, it is reflected back through the light receptor cells and gets a second chance to register. It is this reflective layer of cells which makes it look like your cat's eyes are glowing in the dark.

The great variety and abilities of sight which we see in the different creatures perfectly fits each of their different life styles and needs. Truly, the Lord provides for each creature's unique needs.

Prayer: Dear heavenly Father, You do not ignore the needs of any of Your creatures. Assure me when I lack Your peace, that You have not left me alone, either. In Jesus' Name. Amen.

Mechanical Excellence Under Stress

Psalm 148:7a, 8
"Praise the Lord from the Earth, . . . fire and hail, stormy wind, fulfilling His Word . . . ."

Scripture does not portray God as having created the world and its principles, and then allowing the whole works spin off into space so that He no longer has to pay attention to the details. Scripture tells us how God pays attention to every detail of the creation.

Few living things are like a brick wall. Rather, they move or change shape in response to high speed wind or water movement. Take leaves, for example. As the wind increases, leaves automatically change shape with respect to the wind direction so that they offer less wind resistance — in other words, they become more streamlined. Holly leaves, for example, fold up and over each other along the branch, cutting the area exposed to the wind by over 50 percent. The same is true of an entire tree. As the old saying goes, the tree that bends with the wind is less likely to break. So the tree bends, lowering its resistance to the wind.

Even people, when faced with a strong wind, walk and move differently than in a light breeze. Recent scientific studies on the amount of drag which living things have has shown that living things generally offer less than half as much drag as non-streamlined objects like buildings. In fact, living things have 30 percent less drag than modern, streamlined automobiles.

God has not missed any details in His design of the creation. Modern technology has yet to come up with better designs, even in this simple area of engineering. Nothing has been left to chance!

Prayer: Dear heavenly Father, I am thankful that there is no such thing as chance. Help me to remember that You remain as involved in the creation as ever. In Jesus' Name. Amen.

Your Body's Self-Repair

James 5:16
" . . . and pray for one another, that you may be healed. The effective, fervent prayer of a righteous man avails much."

One of our goals on *Creation Moments* is to invite you to take a couple of minutes out of your busy day to look at yet one more wonderful way in which God's glory is evident in the creation.

Today let's take a moment to look at your body's incredible system for repairing itself. Let's say that you are slicing a potato and — oops — you slip and cut yourself. Almost instantly, a series of precisely ordered steps begins to repair your finger. First the bleeding must be stopped. While the scab is forming over the surface of the wound, the blood below is making another kind of clot out of blood platelets and protein. With the bleeding stopped, your body increases the flow of blood enriched with white blood cells. These not only search out and kill germs, but they clean the wound of damaged cellular tissue. Skin cells begin to increase the rate at which they make new cells in order to bridge the cut with new skin. Underneath, cells called fibroblasts fill the wound to strengthen the tender new tissue, and contract to pull the wound closed. Now blood vessels and nerves complete their repairs as the fibroblasts position themselves along the lines of stress to prevent future damage.

The intelligence in the order in which the steps to healing take place, as well as the advanced biochemistry involved in making those steps happen, makes the healing of a cut finger no less of a miracle than Jesus' healing of the high priest's servant's ear. Healing is clearly His doing, whether it happens slowly or instantly!

Prayer: Dear Father, I confess that I am often too busy to notice how evident Your working is around me. Open my eyes to Your activity and increase my faith. In Jesus' Name. Amen.

Eyes in the Back of Their Backs

Proverbs 15:3
"The eyes of the Lord are in every place, keeping watch on the evil and the good."

While growing up, many of us had mothers who were so good at knowing what we were doing that we often wondered if they had eyes in the back of their heads. While science has yet to find extra eyes in human mothers, it turns out that a species of shrimp literally has eyes in its back.

The shrimp goes only by its Latin name — *exoculata* for short. It lives beneath two miles of water in the mid-Atlantic Ridge, near geological formations called black smokers. Black smokers are chimneys in the ocean floor which continuously shoot out thick black clouds of water, at a temperature of 660 degrees F. No trace of sunlight ever penetrates to such depths. So why do these shrimp need eyes at all? The search for the answer led scientists to discover a source of light they never suspected. They learned that the super-heated water coming out of black smokers is so hot that it actually glows. The glow isn't all that bright, which is why *exoculata* needs such large eyes that they will fit only on the shrimp's back. And what do the shrimp see? They make their living within inches of the hot stream of water. If they get too close to the super-heated water they could become instant shrimp cocktail!

Nothing escapes the Creator's notice. Not these shrimp, nor a life that is in trouble due to difficulty or sin. But we need to remember both His genius in solving problems and that His mind toward us is one of love because of the sacrifice of His Son for us.

Prayer: Dear heavenly Father, there is no problem that is too hard for You. Because I trust in Your forgiving love to me through Jesus Christ, I give all my problems and needs to You. In Jesus' Name. Amen.

The Solar-Powered Pump

Psalm 104:16
"The trees of the Lord are full of sap, the cedars of Lebanon which He planted...."

It may be that one of the reasons that we so often fail to see the Lord's Hand in the creation is that as mighty and wise as His work is, it is usually understated, like the still small voice with which He came to Elijah. Science, stripped of the myth of evolution, helps us to see anew the wonder of His working.

On a nice warm summer day a large tree may pump over a thousand gallons — that's four tons — of water from the ground up to its leaves. The water is collected by the roots, a drop at a time. But the real work of pumping tons of water 30, 60 or 100 feet into the air comes from the top of the tree. As water is pulled toward the tree top, it passes through vessels which have negative pressure in them, pulling the water up. If you were to cut one of these vessels you would actually hear a hissing sound as air rushed into the vessel. Negative pressure in these vessels, high in a tree, has been measured as low as negative 20 atmospheres. This very low pressure is created as water evaporates from the leaves of the tree, creating a vacant space in the vessel which must be filled with more water from below. The genius of the system is that this silent, powerful pump can always deliver as much water as needed.

The engineering excellence and power of the silent pump which delivers water within the tree is a witness against chance evolution, and for our Creator God!

Prayer: Dear Father in heaven, surely even the unnoticed things of the creation bear witness to You. Help me to be a witness of Your forgiveness for us through Your Son, Jesus Christ. In His Name. Amen.

A New Recipe for Primordial Soup

Acts 17:25
"Nor is He worshiped with men's hands, as though He needed anything, since He gives to all life, breath, and all things."

One of the most difficult problems for those who want to explain where everything came from without reference to God is the problem of how life started.

As science has learned more about how complex even the simplest molecules of life really are, scenarios of how life could have started without God became more and more exotic. The most recent of these explanations said that the early Earth had a primordial atmosphere made of ammonia, methane, hydrogen, and water. It was important that there be no oxygen, since oxygen pretty much ruins the chemical reactions which are needed to form even simple biological molecules. It was also important that this early Earth be protected from ultraviolet radiation, which also ruins the chemistry. But this picture of the early Earth has been shown to be totally inaccurate. Recently, a world-wide exhaustive study of the oldest rocks showed that the Earth has always had plenty of oxygen in its atmosphere. And other evolutionists have pointed out that a younger sun would be turning out as much as 10,000 times more ultraviolet radiation than it does now.

Modern science is finding out that what the Bible says is true. Scientists just won't admit it. The Earth has always had oxygen, since life needing oxygen has been around from the first week that the Earth existed. And life can only be explained as the work of the Source and Author of life, our Creator God.

Prayer: Dear Father, I thank You that as the worldly wise are confused by their own false wisdom, You have made the Wisdom of the Ages in Jesus Christ available to the simple through the Bible. In Jesus' Name. Amen.

Smart Brain

Romans 12:2
"And do not be conformed to this world, but be transformed by the renewing of your mind, that you may prove what is that good and acceptable and perfect will of God."

The human brain is the most complex arrangement of matter in the universe. And modern brain research is adding to our wonder about this incredible organ.

It was once thought that various behaviors and abilities were locked into position, each in their own parts of the brain. But medical researchers are learning that if the part of the brain that controls your left hand, for example, is damaged beyond repair, other parts of the brain will take over for the injured part. This doesn't always happen automatically, however. It takes effort on your part. If you don't use the hand, the brain won't work very hard to re-assign its function to another part of the brain. It also works the other way. Even if the brain isn't damaged, failure to use your left arm will result in a shrinkage of the brain area which controls it. So pianists, for example, will have a growth in the area of the brain controlling their skills, if they practice.

These discoveries fill us with more wonder over the incredible abilities of the brain. But they also tell us something about ourselves. We are not, as many think, locked into certain bad habits or sins. We are learning that the transformation of our minds by God's power is not only something that happens spiritually — physical changes take place in our brains as a result of our behavior. Is your mind being transformed by the power of God, or is it being conformed to the world?

Prayer: Dear heavenly Father, I thank You for the wonderful gifts You have given me. Transform me by the power of Your Word so that I may be more like my Savior, Jesus Christ. In His Name. Amen.

Retracting Mosquito Antennae

Psalm 40:5a
"Many, O Lord my God, are Your wonderful works which You have done...."

All of us are familiar with the drone of the mosquito. It is in the male mosquito that we find one of God's most creative uses of a water-powered engine.

Most people are aware that it is only the female mosquito which feeds on blood. The male feeds on plant juices. It is also easy to see the difference between males and females in most species. Besides the fact that the male isn't lining up to bite your neck, it has distinctive antennae. While the female mosquito's antennae are difficult to see, the male's looks like a pair of branched feathers coming out of its head. And if it weren't for a very special feature, these large, feathery antennae would make it difficult for him to fly.

Each antenna is planted in a socket, next to which there is a pad, made out of special protein. This pad is actually a water-powered engine. When flying, the mosquito's antennae are flattened against its head. But when he lands, he raises the antennae so that he can hear. To raise the antennae, a small amount of water from his system is pumped into the pad which increases its size by 25 percent and causes the pad to unfold, raising the antennae.

Nature is full of so many wonders that it would be easy for us to get lost in them. But every one of these wonders is designed by God to lead us to desire to learn more about Him — especially to learn from His Word that He wants a relationship with you through His Son, Jesus Christ.

Prayer: Dear heavenly Father, fill me with wonder and thanksgiving for all Your marvelous works, and especially for Your forgiveness for me through Jesus Christ. In His Name. Amen.

Growing Backbone

Job 38:3b
"... I will question you, and you shall answer Me."

There are many in this world who would deny God as Creator so that they can feel that they need not concern themselves with a personal relationship with Him. However, God makes it hard for people to ignore Him.

Modern medicine has increasingly returned to studying how nature accomplishes medical miracles to bring the benefits of what they learn to mankind. Man has dreamed for thousands of years of being able to restore a severed spinal cord. Once the nerves running inside our backbones are severed, we permanently lose all use of our limbs below the point of damage.

Medical researchers have learned that the black ghost knife fish, native to South America, is actually able to regrow its backbone if it is severed. The black ghost knife fish can even regrow the spinal cord within the backbone and the supporting muscle structure! What's more, scientists have identified the layer of cells responsible for this regrowth. And the best news is that human beings also have the same layer of cells. Unfortunately, the layer becomes dormant by the time we reach adulthood. Researchers are currently trying to learn how that layer of cells, with its wonderful abilities, might be stimulated back into action if needed to repair a severed spinal cord.

This shows us that when we Christians talk about the clear evidence of God's intelligence in the universe, we are not engaging in wishful thinking. Even our best scientific researchers recognize the intelligence with which the creation is made and desire to learn from that intelligence!

Prayer: Dear heavenly Father, I thank You that we can learn and benefit from studying how You designed things to work. But help me to never forget that Your most important work was my salvation through Jesus Christ. In His Name. Amen.

The Ingenious Assassin Bug

Romans 1:20
"For since the creation of the world His invisible attributes are clearly seen, being understood by the things that are made, even His eternal power and Godhead, so that they are without excuse."

For thousands of years Christians have been accused of only imagining that there is intelligence and design to be seen in the creation. Yet the intelligence in the design of the creation is so obvious that sometimes even those who believe in evolution are forced to recognize the intelligence which has been so evident to believers.

Dr. Thomas Eisner, of Cornell University, offers just such an example. Though he is no creationist, in a recent *New York Times* interview, Dr. Eisner pointed out that "every single new idea in chemistry has come not from the minds of chemists, but from nature." In other words, even our greatest chemists go to the creation to learn their science. This is really a recognition that nature is not put together mindlessly.

Take the female assassin bug's use of chemistry, for example. The female assassin bug will rub the resin of the camphor plant on her belly until she has a good store of it on her. Then, when she lays her eggs, she carefully coats them with the resin. It took chemists some time to learn that the resin acts in the same way as do moth balls against moths — it keeps ants from eating her eggs. So the female assassin bug has shown chemists a new ant repellent!

The reason that God's Hand is so evident in the creation is that He wants all people everywhere to know that He exists as a personal, intelligent Being. Once that is established in people's minds, He seeks to draw them to a personal relationship with Himself through His Son Jesus Christ.

Prayer: Dear heavenly Father, I give thanks to You that You have sought me through the gospel and worked saving faith in Jesus Christ within my heart and mind. In Jesus' Name. Amen.

Mysteries of Breaking Glass

Isaiah 43:4

"All the host of heaven shall be dissolved, and the heavens shall be rolled up like a scroll; all their host shall fall down as the leaf falls from the vine, and as fruit falling from a fig tree."

Everything in the world is running down — nothing will last forever. We encounter examples of this every day. It would be very depressing if we didn't know the reason for it, and what God has done about it.

Take the simple example of glass. Newly-weds don't have to be married very long before their only memory of that set of everyday glasses they received is held in the one, remaining, unbroken member of the set. Yet, if glass were perfect, it would be much stronger than steel. But no glass is perfect. Its surface is full of tiny cracks, too small for the human eye to see. These cracks slowly creep along the glass, completely invisible to the naked eye. This is why a glass can be dropped several times and not break. And then, one day as you are picking it up, it shatters into a million pieces. Glass "remembers" every stress it receives. If you jar the glass, the cracks can grow quickly — as fast as 60 miles per hour!

Most amazing is that this behavior is so strange, considering that glass is not really a solid. Glass never truly solidifies from a thick liquid state. Sometimes you can even see the sag caused by the flowing glass on window panes that are over a century old.

We encounter the natural degeneration of the world every day. As St. Paul writes, the entire creation groans under the consequences of sin. But thank God that He refused to leave us subject to decay and death — for He sent His only Son to save us from every consequence of sin. In Christ there is no more death and decay!

Prayer: Dear Father in heaven, I thank You that we have not been left to be victims of the degeneration in the world because of sin. I pray that for the sake of Your Son Jesus Christ You would help me to always remember this. In His Name. Amen.

Young Fish Learn New Tricks

Proverbs 1:8-9
"My son, hear the instruction of your father, and do not forsake the law of your mother, for they will be graceful ornaments on your head, and chains about your neck."

There is an old saying that all too often proves to be true: very young children think their parents know everything. As young children grow into teenagers, they decide that their parents are the most stupid people God ever made. Later, when they are raising their own children they discover how wise their parents really were! But when our current world has chosen to talk about animals, it talks not about intelligence and learning, but about food supplies and instinct. After all, learning and intelligence are products of supposed human evolution.

However, science is now beginning to discredit this myth of evolution. In a study published near the end of 1988, biologist Robert R. Warner of the University of California, showed that even adult fish teach their offspring the facts of life. Warner's study involved moving fish called wrasses around to various reefs where they prefer to mate. He discovered that if there were no other wrasses at the reef, the young wrasses would establish their own mating sites. But if there were already adult wrasses at the reef, the young wrasses would share the mating sites which were taught to them by the established adults. In other words, the adults taught their traditional mating sites to the next generation!

As the work of an intelligent Creator, the creation and its creatures are filled with intelligence. This teaches us that the intelligence levels of certain creatures has nothing to do with their supposed evolution. It also teaches the younger among us that it is a natural part of God's plan that we learn from the experience of those who are older than ourselves.

Prayer: Dear heavenly Father, I have so much to learn, not only from those around me, but also from You. Help me to desire what You have to teach me in Your Word and to be an eager student. In Jesus' Name. Amen.

A Day at the Flea Circus

1 Samuel 26:20b
"For the king of Israel has come out to seek a flea, as when one hunts a partridge in the mountains."

The flea seems to be the universal symbol of smallness and unimportance. The youngest or smallest among a group of young boys today may be affectionately nicknamed "flea" by his friends. When King Saul pursued David to kill him, David compared himself to a flea by asking whether the king had come out with soldiers to kill a flea. But just because a flea is small, doesn't mean that God treated it as unimportant when He designed it.

The flea, with its ability to jump many times its own length, is a miniature marvel of engineering. If you see a flea jump — and that's not easy to do since they are indeed very small — you will wonder whether the flea is actually powered by springs. The amazing truth is that the flea *is* powered by tiny springs. When it jumps, the flea actually releases more than five-and-a-half times as much energy as the most perfect muscle tissue can generate! The flea is able to do this using normal muscle output because he has small pads of a natural protein rubber called resilin in his legs. As he slowly depresses the pads, he stores the energy which will be released in his next leap in one-seventh the time he took to store it!

To our Creator, nothing and no one is unimportant. No one is a "flea," so to speak. This is because He is, by nature, a loving and caring Creator. Do you know His love? Do you know that He seeks a personal relationship with you through what Jesus Christ did for you on the cross? Learn more about His love for you in the pages of Holy Scripture today!

Prayer: Dear Father in heaven, it was Your Son, my Savior Jesus Christ, Who taught me to call You "Father." Help me not to take Your love for granted, but always to thank You for Your love for me. In Jesus' Name. Amen.

Warming to an Old Climate

Genesis 2:6
". . . but a mist went up from the earth and watered the face of the ground."

It was the largest bird that ever lived. It stood ten- to twelve-feet tall and could easily out run a horse. Being a meat-eater, this fossil bird, which has been nicknamed "terror bird," possibly *did* chase horses and bring them down with its sharp talons. The only thing that limited this creature's terrible abilities was its inability to fly.

"Terror bird," described by one scientist as the most dangerous bird to ever live, existed in a world which also included other large birds — one had a wing spread of 17 feet! In the warm climate — not unlike a greenhouse — ferns towered 150 feet above the tropical forest. Huge alligators, marsupials, and a strange mix of other creatures hid within the forest of fig and sequoia. Where would one find such a place? Of all places, Antarctica! Yes, it was a long time ago — although not as long ago as evolutionists claim. But these fossil creatures are only a few of the many evidences that the Earth was once much warmer. Likewise, dinosaurs and other evidences of a once-tropical climate have also been found in Siberia, Alaska and the Arctic.

It sounds as if a tropical forest is being described when Genesis talks about mist rising from the ground, watering the plants. If the Earth really *is* warming up, maybe it's just now returning to normal. This is certainly not something which the ancients in the Near East would have made up — offering further evidence of God's authorship of Scripture!

Prayer: Dear Lord Jesus Christ, You are the Word made flesh for my salvation, for which I thank You. Draw me closer to Yourself through Your Word of Scripture so that I may be even closer to You. Amen.

The Helpful Honeyguide

Job 35:10-11
"But no one says, 'Where is God my Maker, Who gives songs in the night, Who teaches us more than the beasts of the earth, and makes us wiser than the birds of heaven?'"

It has been known for years that the African honeyguide leads badgers to bees' nests, where the badger finds the honey that he so likes to eat. When the badger tears a nest apart to get at the honey, he leaves more than enough behind to satisfy the honeyguide.

Now it has been learned that the honeyguide has a similar relationship with the Boran people of Kenya. When the honeyguide has found a bees nest, it will alert the Boran through flight patterns and calls, bidding them to follow the bird to a honey site. If the Boran want to know where honey is, and there appears to be no honeyguide around, the Boran whistle to call the honeyguide. As the honeyguide leads the Boran, it keeps its flights short so that it is always in view. The honeyguide also calls to the people so that they know which way to go. When the Boran reach the honey, they always make sure that they leave some for the honeyguide. Researchers also reported that they saw honeyguides scouting out bees' nests at night so that they had good sites to which to lead the Boran the next day.

While the honeyguide does get its reward of honey in return for its help, the intelligence of the honeyguide in finding honey and establishing these relationships with human beings is impressive. But while the honeyguide can help teach us that the creation is the work of an intelligent Creator, it cannot teach us how to have a relationship with Him. For that we must go to the Bible.

Prayer: Dear Lord Jesus Christ, all things were made through You, and for that we give thanks to You. Open the eyes of those who do not know You so that they might see Your Hand in creation, and so be led to find Your love in Scripture. Amen.

Frozen Turtles

Job 38:28-29
"Has the rain a father? Or who has begotten the drops of dew? From whose womb comes the ice? And the frost of heaven, who gives it birth?"

Many creation scientists believe that one of the reasons why there are few, if any, dinosaurs left in the world today is that the world after the Flood is a cooler, wetter place than most reptiles like. However, there are exceptions to every rule. Turtles, which are, of course, reptiles, live in places like the northern United States where temperatures drop far below freezing. How do the turtles stand it?

Some creatures make it through cold winters by hibernating. Others burrow deeply enough into the ground to avoid freezing temperatures. But new scientific research has shown that at least some turtles actually freeze during the winter. Studies on the painted turtle, common in the northern United States, show that this species can survive being frozen. The turtles can actually be in water which freezes solid. As long as something less than 54 percent of the water in their bodies freezes solid, they can thaw out and survive quite nicely. It appears that the turtle has a couple of strategies which allow it to get away with this. Blood sugar levels in the turtles nearly triple when they are frozen, and there is a sharp increase in certain amino acids which act as antifreeze. In addition, glycerol, another antifreeze, triples. Researchers point out that while these changes supply some answers, they don't completely explain the turtle's survival.

Scientists are trying to find out just how painted turtles live through being frozen. They hope to learn how to preserve human organs longer for later transplantation. In effect, they are trying to find out how God enables the painted turtle to survive freezing so they can copy His method.

Prayer: Dear wise heavenly Father, I stand in amazement at the way in which You have created such a variety of creatures and provided for them. Help me to understand that You care for me even more than them. In Jesus' Name. Amen.

Blister Beetle Protection

Psalm 7:1
"O Lord my God, in You I put my trust; save me from all those who persecute me; and deliver me."

Everyone, and it seems nearly everything in the creation, wants to be safe. If threatened, we want to have protection. But most of us, and most living things, prefer to be ignored by danger, even if ignoring danger doesn't make it go away. Unfortunately, many people do not see that the similarity stops at this point. Human beings face many more threats than do animals.

Many of the methods used for self-protection in the animal world show a great amount of creativity and intelligence. For example, the blister beetle oozes a chemical which effectively deters ants and other predators that might disturb it. The beetle's entire body serves as a trigger for this self-protection mechanism. As soon as any body part — say a leg — is disturbed, that part of the body begins to ooze the nasty chemical, driving off the predator.

It is surely true that appeal is in the eye of the beholder, for the male blister beetle uses this same chemical to attract female blister beetles! The ornatrix moth employs the same chemical to attract female moths. In fact, many of the chemical tricks used by man for self-protection were first used by insects and other animals. Again, man learns from the way the Creator has designed things.

But the threats against us human beings are much greater. Sin, death, and the devil are very real forces in our lives, and without the protection of Jesus Christ would continue to be constant dangers into eternity. But thank God that He has provided protection from *all* the dangers we face through the work of His Son, Jesus Christ!

Prayer: Dear heavenly Father, I thank You that You are my protector and that Your Son, Jesus Christ, has done everything necessary to save me from sin, death and the devil. Help me not to treat this gift lightly. In Jesus' Name. Amen.

Natural Fire Insurance

Job 12:7-9
"But now ask the beasts, and they will teach you; and the birds of the air, and they will tell you; or speak to the earth and it will teach you; And the fish of the sea will explain to you. Who among all these does not know that the Hand of the Lord has done this...?"

The 1988 fires in Yellowstone National Park were the worst ever recorded there. Smoke from the fires was visible over much of the United States and Canada. Scientists have been studying the fires from all angles, and they are coming up with some surprising conclusions.

Researchers now say that it is normal for such huge fires to occur every couple of centuries in Yellowstone. And while fires burned about 20 percent of the park, less than 1 percent was so devastated by fire that the soil was made lifeless. Further, periodic fires speed the release of nutrients to the soil that are trapped in dead wood. Many plants actually depend on fires to shock them out of an unproductive dormant period. The lodgepole pine, which makes up 77 percent of Yellowstone's forests, actually has its own fire insurance. The pine produces two kinds of seed-bearing cones. One cone releases its seeds normally. But the other cone is coated with a strong resin which keeps the cone sealed shut. These cones may remain sealed shut on the tree for decades until a fire burns off the resin and causes the cone to release its seeds. After a fire, there are up to 20 seeds for every square foot of forest lying on the ground, ready to sprout in the newly-enriched soil!

If the lodgepole pine could talk in words we could understand, it would tell us of the wisdom and love of the Creator so wisely providing for its needs. But even though they cannot speak with human words, they have been able to communicate the wisdom of this arrangement even to amazed but unbelieving scientists!

Prayer: Dear Lord Jesus Christ, King of creation, help me to join, with my human voice, all of Your creatures who bear witness to Your wisdom and praise You with the voices You have given us. Amen.

Bird School

Isaiah 16:2
"For it shall be as a wandering bird thrown out of the nest; so shall the daughters of Moab at the fords of the Arnon."

We human beings are not the only ones who compare losing our homes to the orphaning of young birds. The Bible does, too. And it is this fact that some Bible critics use to illustrate their claim that the Bible is nothing more than a humanly-devised book. Sure, we know that birds usually take very good care of the material needs of their young. But there is much more to it than that, as Scripture suggests — and science now confirms.

It used to be thought that all animal behavior was instinctive. After all, many scientists said, animals are so much lower on the evolutionary scale than human beings. Now scientists have learned that at least some birds, and most likely most if not all birds, *teach* their distinctive song to their youngsters. In one recent study, young cowbirds were paired with only songless female cowbirds from another part of the country where the cowbird song is distinctly different. Before long, the young cowbirds had totally reworked their songs to match the area of the country from which the females came, even though the females had never uttered one note! Further research showed that the females taught the new singing style to the males using only motion and touching — very sophisticated communication indeed!

Clearly, it is pride which leads man to say that the Word of his Creator is in error. But pride does not only belong to the unbeliever — it is part of the old Adam in all of us. That is why we need the constant instruction of our heavenly Father Who Himself teaches us with more than human words!

Prayer: Dear Father, please do not allow me to live like an orphan without a heavenly Father when I have such a loving and caring heavenly Father as You! Take away my pride and make me hungry to learn all that You would teach me. Amen.

God's Water Engine

Genesis 1:2b
"And the Spirit of God was hovering over the face of the waters."

Water is one of the most basic and powerful substances that God created. So it is little wonder that even on the first day of creation, even before there was light, water is mentioned. We have, in other *Creation Moments*, talked about the strange ways in which water acts. Today, let's talk about the tremendous *power* in water.

Perhaps the first thing that you think of when the power of water is mentioned is the way in which water has cut and eroded the rocks of the Earth. But that is just kid's stuff compared to what I have in mind. Some of the most impressive power of water is seen when it powers one of the water engines which God has created. You see, there are certain substances, like the jelly mass in which frogs' eggs are suspended, which are hungry to absorb immense amounts of water. After a short amount of time in the water, a swollen mass of frog eggs may be larger than the mother frog from which they came. And this engine does its most impressive work when it is enclosed so that it cannot easily expand as it absorbs water. The resulting pressures have been measured at thousands of times higher than atmospheric pressure. In practical application, a small number of dry bean seeds, accidentally left under a concrete sidewalk, will, when they get wet, swell with such power that they will break the concrete!

God uses seemingly simple and powerless things to do incredible feats. His Word, even though it takes the same form as human words, creates worlds and galaxies. But more amazingly, and more powerfully, it changes human hearts and minds!

Prayer: Dear Lord Jesus Christ, let me be reminded often by Your creation that You do what is impressive using those things which seem common and unimpressive. Do not let me neglect the simple power of your Word in my life. Amen.

Reptilian Fuzzy-Feet

Psalm 101:3

"I will set nothing wicked before my eyes; I hate the work of those who fall away; it shall not cling to me."

It is the noisiest reptile on Earth, and the smallest, though it is related to the dinosaurs. It can ignore gravity, walking up walls and across ceilings. Only in Java is it considered good luck. But in many parts of the world, despite the fact that it is harmless, and clears homes of insects, it is considered repulsive or even demonic. Yet the tiny gecko measures only an inch-and-a-half long if you include its tail.

There are 670 species of geckos around the world. They are found in deserts and jungles, and they like living in homes where they are excellent at controlling insects, while leaving almost no trace of their presence. But the feature of the gecko which inspires the most wonder is its ability to effortlessly walk up a wall or across the ceiling. To accomplish this neat trick the gecko has pads on his feet. Under the microscope these pads look like tiny pin cushions. Each microscopic bristle contains even smaller branches, enabling the gecko's foot to hook, with these microscopic hooks, into the smallest irregularities. Some geckos have over a billion of these hook-like hairs on their feet! Incidentally, this explains why geckos walk with that funny skitter. Each foot must be unhooked before it is lifted for the next step.

Unfortunately, when the gecko is stuck to a wall, and the wall falls, he will fall with it. The same is true of us. If we cling to evil — we will end up going where the evil goes. It is not by our own good intentions, but only through Jesus Christ that we are released from evil through the forgiveness of sins. Look to Him!

Prayer: Dear Lord Jesus Christ, make me more uncomfortable when evil comes before my eyes. Do not let evil cling to me, but wash me clean in Your blood, and give me Your peace. Amen.

What Anemones Hear

Genesis 1:21a, 22a
"So God created the great sea creatures and every living thing that moves, with which the waters abounded, according to their kind. . . . And God blessed them, saying, 'Be fruitful and multiply, and fill the waters in the seas . . . .'"

While it's pretty easy to find an elephant's ear, its not at all easy to find a frog's ear. But how would you like to be one of the scientists who had to look for a sea anemone's ear!

God knows of and has made far more than our imaginations could ever dream of. The first scientists of the modern era knew this; they described their work as "thinking God's thoughts after Him." We need to be open to discover God's wonderful flights of imagination. So while the evolutionist is trying to fit creatures into his simple-to-complex scheme of things, those of us who believe in the Creator are open to wonder at things that evolutionists might never discover. The recent discovery that the supposedly very "simple" sea anemone has a more complex "ear" than any other creature, is one such example. In fact, scientists remain mystified at how the anemone can hear certain sounds which are beyond any other known creature's hearing. The discovery that anemones can hear was released only recently by scientists who were studying how anemones knew when to fire their microscopic, poisonous darts to kill their prey. They found that anemones not only hear, they can hear a huge range of frequencies by doing something no other creature can do; change the frequency range that their hearing receptors — which are similar to ours — can pick up.

Our Creator is a God Who communicates. The Bible tells how He communicated with each of His creatures at the time of creation. But the Bible is His communication to you and me this very day!

Prayer: Dear Father, I confess that very often my ears are not able to hear Your Word because they are filled with other sounds as well as cares and worries. Please help me to better hear Your Word and know Your Son, Jesus Christ. Amen.

The Fossilized Hats

Luke 19:40
"But He answered and said to them, 'I tell you that if these should keep silent, the stones would immediately cry out.'"

How long does it take to make a fossil? Simply because the process of changing something into stone seems so strange, most people think that it takes a long time to fossilize something. So, many find the millions of years that evolutionists talk about to be very believable.

But experience — which is, after all what science is all about — shows that it is not unusual for things to become stone very quickly under the proper conditions. The process of fossilization, where minerals replace the original material, is illustrated very nicely by a lost miner's hat. A miner lost his hat in a mine in Australia, and it was found 50 years later, completely turned to limestone. The hat is now in a mining museum in Tasmania. Even more dramatic is the bowler hat that was buried in a volcanic eruption in 1886 in New Zealand. When it was excavated — only 20 years later — it had already been turned to stone. Even more surprisingly, a leg of ham which was buried in the same eruption had also been fossilized in just a few short years! Imagine — if these items would have been found even a thousand years later, when no one remembered the eruption or that anyone had ever lived there, these items *might have been declared millions of years old!*

Many, if not most, of the fossils which are found today were created during the great Flood of Noah's time when billions of plants and animals were quickly covered by water. They bear witness to God's judgment of sin. These modern fossils also bear witness against the idea that we are the product of millions of years of evolution!

Prayer: Dear Lord Jesus Christ, it is true that even the rocks bear witness to You as the Creating Word. I pray that this may not be true because Your people do not bear witness to You today. Help me to be a better witness for You. Amen.

Did Life Start in the Sea?

Psalm 146:5-6
"Happy is he who has the God of Jacob for his help, whose help is in the Lord his God, Who made heaven and earth, the sea and all that is in them; Who keeps the truth forever."

Many people find the sea a mysterious place that they have never visited. It is true that the sea is the last, most unexplored frontier left on Earth. For these reasons the ancients made up stories about the sea, and many of which reflect their natural fear of the unknown. One of the most wide-spread *modern* myths about the sea is that life began there.

All of the elements found on Earth are present in sea water. Evolutionists point out that there is a similarity between the salt contents of sea water and blood. This is evidence, they claim, that life began in the sea. While it's true that there is a similarity between the salt contents of sea water and blood, the reason for it has nothing to do with evolution. You see, both sea water and blood are water-based solutions. Since all water-based solutions must follow the same principles, simple chemistry dictates some universal similarity for all water-based solutions. All that is needed, in addition, is the presence of salt. Since salt is necessary for life, blood is forced, by simple chemistry, to be somewhat similar to *any* other water-based salt solution. So you see, the similarity between sea water and blood has nothing to do with evolution. Besides, many of the elements in sea water are poisonous to living systems, although they are not present in enough concentration to hurt us. If life developed among these elements it should not find them at all poisonous.

It is not to the sea that we should look for the source of life. On this point, both the Bible and true science agree!

Prayer: Dear Lord Jesus Christ, I thank You that it is through You that I and all living things were made, and not through the impersonal, unloving sea. Enable me to communicate this truth in our troubled world. Amen.

Air-Cooled Elephants

Genesis 2:1-2
"Thus the heavens and the Earth, and all the host of them, were finished. And on the seventh day God ended His work which He had done, and He rested on the seventh day from all His work which He had done."

Every engineer will agree: if a particular design does a good job, every time that particular need arises, the wise engineer will re-use the design. After all, why re-invent the wheel every time you design a new car? On the other hand, if there is no intelligence guiding the solution to a design problem, one would expect to find many inadequate design solutions to the same problem.

Large, warm-blooded creatures, because of their mass of heat generating cells, have the problem of generating what could be, in warm climates, dangerous amounts of heat. What then, would be an intelligent design for something as large as an elephant which is supposed to live in a warm jungle climate? Well, first of all, you would give the elephant a slower metabolism, so that he does not generate so much heat in the first place. But if you give him too slow a metabolism, the elephant will not be able to live. So, as a wise engineer, you design a cooling system for your elephant. The fact is, the elephant's ears, weighing over 100 pounds each, are indeed cooling devices, filled with many small blood vessels. By changing how close its ears are held to its body, or flapped, the elephant can control how much the blood in its ears is cooled before it is returned to the rest of its body. As you can see, the elephant's cooling system is remarkably similar to that of a car.

By the time the Lord finished the work of creation, He had to solve millions of engineering problems just like this. And these are often the same problems that modern engineers must solve. Is it any wonder, then, that many engineers are creationists!

Prayer: Dear Father in heaven, help me to have joy in my work, as You felt joy in Your work of creation. Take all that I do, whatever it is, as part of my offering for You. In Jesus' Name. Amen.

Silken Traps

Deuteronomy 12:30
"'Take heed to yourself that you are not ensnared to follow them, after they are destroyed from before you, and that you do not inquire after their gods, saying, "How did these nations serve their gods? I will also do likewise."'"

While spider webs can be beautiful things, most people do not find most spiders all that attractive. Some spiders are deadly, even to humans, and many of them are capable of giving nasty bites. So in some cases danger lurks with beauty.

While some caterpillars, centipedes, worms, and scorpions make silk, spiders are the only creatures which depend on silk for catching prey. There are many species of spiders. Depending on which species you are talking about, spiders have up to seven silk making glands which produce a large number of different kinds of silk for different purposes. Silk from one gland may be used to wrap prey, while silk from another gland may be used to spin scaffolding. The female spider even has a special silk she uses for covering her eggs. The silk flows to the spinnerets out of each gland as a liquid. These organs, on the rear part of the spider's abdomen, are movable and have nozzles at the end. The nozzles on the spinnerets can be used alone, or combined in order to make a blend of different silks. The spider then draws the silk from the spinneret. The faster the spider draws the silk, the thinner and stronger it is. As you can see, the spider has been given a lot of versatility in making silk.

Deadly snares for God's people can be attractive — even exquisite in design. The devil knows that we will not be snared by those things which are not attractive to us. Yet when we are snared and fall, we must remember that we have a Savior Who has already overcome the snares of the devil in our place. And in Him there is forgiveness!

Prayer: Dear Lord Jesus Christ, I thank You that You have overcome sin, death and the devil in my place. I pray with the disciples: Lord I believe, help my unbelief, which the devil uses to snare me with attractive lures. Amen.

Where Do *Creation Moments* Come From?

Hebrews 11:13
"These all died in faith, not having received the promises, but having seen them afar off were assured of them, embraced them, and confessed that they were strangers and pilgrims on earth."

One of the most frequent questions we at Bible-Science are asked as a result of *Creation Moments* is: Where do you get all that fascinating information? Is there a book where all these interesting facts are assembled together?

In a very real sense there is a book of sorts where our information can be found, besides the book in which these scripts are available. That book is God's creation. Many people have written to tell us they have gained a deeper appreciation of God's work of creation through *Creation Moments*. I have to say that I have, too. And there is the key. Six years ago when I first started researching for material for these broadcasts, it often took me some time to find the subjects and materials we have been airing. Now, after doing this for several years, I find that I run into good information for *Creation Moments* without even looking! It's not that there is more information available now than there was before. Rather, my vision in seeing the Hand of our Creator has increased. And there's the key to how any of us find this kind of information. We *learn* to see it. The idea for a broadcast may come from a magazine article or from some animal profiled on a nature program or a newspaper article. The Hand of our Creator is all around us. We just need to learn to have better vision so that we can see it.

There's one other key to this vision. It certainly helps us to get used to looking at all things in this world through the eyes of faith, and that means being tied in to the Word of God on a regular basis. What we talk about on *Creation Moments* is all around us for the seeing. I hope and pray that we have helped you see a little better!

Prayer: Dear Lord Jesus Christ, I thank You that through faith, created by Your Word, I am able to see wonderful things which are unseen by the wise and proud of the world. Through Your Word build my faith and increase my vision. Amen.

The Lighter Side of the Bible

Matthew 23:24
"Blind guides, who strain out a gnat and swallow a camel!"

Some Christians give others the idea that Christianity is a dull and joyless faith, and that God is stern and humorless. But any Creator Who purposely makes the playful kitten or the silly monkey certainly has an appreciation for humor in its proper place. The Bible itself is not short on puns and humorous word plays, either. But the printing history of the Bible offers its own humor — in this case, based on human error.

For example, did you ever hear of the **Bug Bible**? Published in 1551, the **Bug Bible** got the name when, for some unexplainable reason, "Thou shalt not be afraid for the terror by night" from Psalm 91:5 was printed as, "Thou shalt not be afraid of any buggies by night." Human error was also presumably responsible for **The Wicked Bible**, published in 1717. The **Wicked Bible** got its name because the word "not" was *left out* of one of the Commandments, leaving the Commandment to read, "Thou shalt commit adultery."

One of the most amusing word plays by our Lord is found in Matthew 23:24, where He warns hypocrites about straining at gnats and swallowing camels. In Aramaic, the language Jesus spoke, the word for gnat is nearly identical to the word for camel. The verbal humor, which is an excellent way of teaching, produced a simple slogan that was easy for the common man to remember: "Blind guides, who strain out a *galma*, and swallow a *gamla!*"

Despite occasional human printing errors, we can be sure that the Bible is the inerrant Word of God. And despite God's sincere earnestness in calling us to Himself, we can also be sure that He created humor.

Prayer: Dear Father in heaven, through Your Word of promise fill me with true joy and a lightness of heart so that others may see that being Yours is a joyful, wonderful thing to be desired. In Jesus' Name. Amen.

Nature is the Best Chemist

2 Chronicles 16:12
"And in the thirty-ninth year of his reign, Asa became diseased in his feet, and his malady was very severe; yet in his disease he did not seek the Lord, but the physicians."

Aristotle is quoted as once saying that the main job of the physician is to amuse the patient while nature heals him. Of course, it is true that modern medicine has greatly benefited man. Nevertheless, sometimes even the greatest physician can do very little but make the patient comfortable. And sometimes the modern doctor can still learn from the medicine man.

A couple of years ago, a chemist from the University of California was visiting villages in the East African Bush. There he witnessed entire tribes lining up to drink a tea made by the medicine man from the berries of a local wild bush. It was said that the tea would ward off cholera. A little local study revealed that the tea did indeed seem to prevent cholera. So the chemist collected a quantity of the berries and took them back to his lab for study. He found that the active ingredient in the berry, maesanin, does act as a powerful antibiotic. Yet it doesn't cause the body to produce antibodies as do other antibiotics, nor does it fight bacteria in the same way as other antibiotics. This means that maesanin could be effective for people who are sensitive to other antibiotics. The chemist concluded, and we quote, "Nature is still the best chemist."

Of course we know that the so-called "nature" is not an impersonal set of forces, but the Creator Himself. He created the bush and the berries that offer what appears to be an antibiotic which could be superior to any before discovered. And He freely gave it to us. Yes, in some cases, doctors can do us a lot of good, but in every case God is ready, willing, and able to help, for the sake of the saving work of His Son on our behalf.

Prayer: Dear Lord Jesus Christ, You are the Great Physician of both body and soul. Help me always to remember to look to You first when I am in need — and to thank and praise You when I am not in need. Amen.

Does the Bible Err on *Pi*?

1 Kings 7:23
"Then he made the Sea of cast bronze, ten cubits from one brim to the other; it was completely round. Its height was five cubits, and a line of thirty cubits measured it circumference."

Everyone knows that when you divide the circumference of a circle by the diameter, you get three and one-seventh, the value of *pi*. Even the ancients knew about *pi*. But how is it then, that 1 Kings 7:23 tells us that the bronze Sea which Hiram made for King Solomon was thirty cubits in circumference but ten cubits in diameter? Those measurements don't yield a value of three and one-seventh — *pi*. Is this a mistake in the Bible?

For thousands of years Bible critics have pointed to 1 Kings 7:23 as an error in the Bible. They support their claim that the Bible is the product of man's imagination with this passage, saying that the ancient Jews evidently didn't even know about *pi*. Even though this challenge was answered by a rabbi in the second century A.D., critics still bring it up today. The solution to the problem is really quite simple. The circumference of the vessel is measured from the *inside* walls of the vessel, while the diameter is measured from the *outside* walls. The thickness of the walls of the vessel makes up the missing "one-seventh" value of *pi*! When confronted with this explanation, even many modern critics find it reasonable.

We can thank the Lord that He has given us His Word, without error. He knew that the last thing mankind needed was another book with human error in it. So no matter how much man and his learning exalts itself against God, His Word proves true. And it does so for a very good reason: because it is in the pages of the Bible that we learn of the saving work of Jesus Christ for us!

Prayer: Dear Lord Jesus Christ, You are the Word made flesh for my salvation. I thank You for the reliability of the Bible You have given us. I pray that You would fill me with a burning desire to study Your Word more than I do now. Amen.

The Reluctant Fish

Habakkuk 1:14
"Why do You make men like fish of the sea, like creeping things that have no ruler over them?"

One of the strangest creatures in the world builds mud walls to protect its territory. This creature, a fish, digs two burrows in the mud, up to two miles apart, across land, and travels from one to the other. And even though it is a fish, it does its mating dance, which can last up to an hour, on land. It almost seems as though the mudskipper is rather reluctant to be a fish.

The mudskipper spends so much time out of the water because it has the ability to absorb oxygen from water on its skin. All it needs to do is stay wet. It also has sponge-like sacks around its gills which hold a supply of water. And should any of these systems not be enough, the mudskipper can also take air in directly. And just as our eyes do not see well under water, most fish cannot see well out of the water. But the mudskipper has special eyes which give him good vision in and out of the water. The mudskipper is a member of the gobi family, found on tropical shorelines from Africa, through Southeast Asia, to Japan and the Philippines. Another member of the gobi family not only leaves the water, but even climbs trees!

The mudskipper is an excellent example of God's unbridled creativity. Imagine giving such a wondrous design to a fish! The prophet Habakkuk compares men to fish when he asks why God allows men to be caught by evil people, as fish are caught by men. But God reminds Habakkuk that when the evil attack the godly, they are storing up punishment for themselves. Ultimately, the Lord tells Habakkuk, and us, that all things serve the glory of God. Or, as the apostle put it, "All things work together for good to those who love God."

Prayer: Dear Father in heaven, forgive me for those times when I worry, forgetting that You are in complete charge and that you have an unlimited mind for finding solutions and bringing good out of the worst situation. In Jesus' Name. Amen.

The Brain's Master Clock

Daniel 2:21
"And He changes the times and the seasons; He removes kings and raises up kings; He gives wisdom to the wise and knowledge to those who have understanding."

Our bodies have many different rhythms. For example, your body temperature fluctuates throughout the day, tied to a cycle of about 24 hours. A person's blood pressure can change by as much as 20 percent, following a 24-hour cycle. Some body rhythms go through their complete cycle in a few hours, others in seven days, and others in about 28 days. In each case, one or more organs in the body control each rhythm.

Some bodily rhythms are controlled through the hypothalamus. Your pituitary controls still other rhythms. Still other organs control other cycles. This arrangement sounds sort of like standing in a room with hundreds of clocks ticking away. And to make matters worse, each clock goes off on its own time — none are set to the same time. It could be a confusing mess.

But now it has been learned that the brain has one master clock that coordinates all the other clocks in the body. This master clock goes by the initials SCN. It was once thought that the hypothalamus did this work. But the hypothalamus only helps the SCN do its work. Together they turn out a bewildering array of chemical signals for the rest of the body so that everything keeps humming along in fine order.

So here we have one more example of the intricate and wise design of our loving Creator, and how He leaves no detail of life to chance. If He can do this, He certainly wants you to bring all the details of your life to Him as well.

Prayer: Dear Lord Jesus Christ, You teach me that My heavenly Father desires an even closer relationship with me, and that He is intimately involved in every detail of His creation. Teach me through Your Word, and help me to believe what You teach me. Amen.

A Special Star

Genesis 1:14-15
"Then God said, 'Let there be lights in the firmament of the heavens to divide the day from the night; and let them be for signs and for seasons, and for days and for years; and let them be for lights in the firmament of the heavens to give light on the Earth;' and it was so."

If the Earth was only one percent closer to the sun, we wouldn't be here to worry about any "greenhouse effect." And if the Earth were only one percent farther from the sun, there would be plenty of snow and ice on Earth, but no one around to ski on it. The distance from the Earth to the sun is very precisely set to allow life to live comfortably here. The chances of that happening without design are nearly impossible, a fact which has made many an evolutionist uncomfortable.

But there is more. Many stars, perhaps most stars, vary much more than our sun does in the amount of energy they give off. Life would not be possible around many of those stars, no matter where a planet might be located. But our Sun varies its energy output by only about one-tenth of a percent.

Modern science is teaching us much more about our special star. What we are learning is leading scientists to talk less and less about how our star is an "average star." It is becoming clear that an average star — though about the same size as our sun — would probably not be such a nice place to have a home planet, if life would even be possible!

Our growing knowledge of the universe is helping us to see that God has indeed given our special star some very unique characteristics. The sun is specially made for life on Earth, just as Genesis tells us. Science is only just now learning what the Bible has always taught!

Prayer: Dear heavenly Father, Your power and excellent workmanship are clearly evident in Your creation of the Sun and its relationship to our planet. Let this testimony make many open to hear about Your love to us in Christ. Amen.

Population and the Age of the Earth

Genesis 5:1-2
"This is the book of the genealogy of Adam. In the day that God created man, He made him in the likeness of God. He created them male and female, and blessed them and called them mankind in the day they were created."

How long have people been living on the Earth? The evolutionist says millions of years. The Bible-believing Christian says, only about 6,000 years. But the answer to this question is amazingly simple.

If we start with only two people, and they have four children which live to have their own children, the second generation has twice as many people: four. Allowing for infant mortality and other human problems which keep the population down, we still find that, on the average, it only takes about 130 years to double the Earth's population. This figure fits into known historical records. And if anything it's a conservative number.

Now, if man's history extends, say two million years into the past, as the evolutionists say, the Earth ought to have a lot more people than it does now. Or else it took 125,000 years, on the average, to double the human population. That doesn't make any sense at all, especially since human historical records show that it has taken 1,000 times *less* time than that during recorded history.

But if we start with eight people, figuring that the population doubles every 130 years, we find that it takes only about 4,000 to 4,500 years to get a population of 1 billion. And that was the Earth's population in the year 1800 — just about 4,200 years after the Flood — through which 8 people were saved to repopulate the Earth!

Prayer: Dear Lord Jesus Christ, it was through You that all things, including us, were made. When we withdrew our love from God and cut ourselves off from Him through sin, You came to our rescue. How can I ever thank You for everything! Amen.

Plant Mathematics

Romans 1:20
"For since the creation of the world His invisible attributes are clearly seen, being understood by the things that are made, even His eternal power and Godhead, so that they are without excuse."

Throughout the centuries people have noticed, whether they believed in a Creator or not, that there is a mathematically precise underlying structure to the universe and everything in it. One everyday example of this precision can be found in plants. Many plants, including the elm or linden trees, grow their leaves, twigs and branches placed exactly half way around the stem from each other. Next in the series are plants, like the beech tree, with leaves placed one-third of a way around the stem from the previous leaves. Third in the series are plants like the oak with leaves placed at two fifths of a turn. Plants like the holly are next at three-eighths, while larches are next at five-thirteenths — and the sequence goes on. Notice the number sequence of these fractions: 1,1,2,3,5,8,13, and so on. Each number is the sum of the two numbers which come just before it in the sequence. This particular mathematical pattern is called the Fibronacci series, and is recognized as a basic mathematical series. Such mathematical precision is not arrived at by accident.

We marvel at how God has stitched the patterns of the material world onto a mathematical fabric which cannot be missed by those who study it. Such mathematical precision can only be the product of power and intelligence, even as Paul says in Romans 1, "What may be known about God is manifest . . . His invisible attributes are clearly seen, being understood by the things that are made." And while they are not yet flocking to the Bible, this is why many scientists are abandoning evolution!

Prayer: Dear Father in heaven, mathematics is among Your excellent and wise creations. Use this creation of Yours to speak to the wise of the world about Your work of creation, and use me to speak of forgiving grace in Jesus Christ. Amen.

Early Man's Written Language

Exodus 17:14a
"Then the Lord said to Moses, 'Write this for a memorial in the book and recount it in the hearing of Joshua....'"

Evolutionary scientists have been amazed by the discovery that the very oldest artifacts left by man indicate the presence of written language *and* mathematical ability. Creationists, on the other hand, are as pleased at the findings because they predicted that it would be possible to discover that man has had language throughout his history on Earth. After all, man was created by the Word of God — Who later took our flesh upon Himself. Before the fall away from God into sin, man was able to speak directly with God.

While we do not accept the inflated evolutionary years — evolutionists now admit that man's written language goes back at least 30,000 years — we do welcome the discovery of facts so supportive of the biblical view of history. When archaeologist Alexander Marshack determined that one ancient carving was a calendar, scientists sat up and took notice. This discovery not only meant that ancient man had a sense of time, but that he also had a written language and the desire and ability to do math. Other evidences of written language have since been studied and, to some extent, deciphered. But what they show is that no matter how far back one is able to find evidence of man's work, language was already established.

Man is man and has always been man. If our actions are less than human, it is because sin has taken over. But there is rescue from sin in Jesus Christ, our Savior from sin and its effects. If you don't know Him, or if your life seems less than human, take a look at the Bible and see what He has to offer you.

Prayer: Dear Lord Jesus Christ, You are the Word made flesh for our salvation. Help Your people see, especially in this doubting world, that just as You are perfect, so is Your written Word perfect, given to us so that we might know of Your love. Amen.

Ant Genius

Proverbs 6:6
"Go to the ant, you sluggard! Consider her ways and be wise. . . ."

Ants seem to know what they want to get done, and how to do it. While we have all noticed ants' "can do" attitude — remember that old rubber tree plant — one scientist decided to find out just how clever ants are.

The setting was the scientist's weekend cottage in the country. The scientist placed a saucer of chocolate on a wooden stool inside a tub. The tub was partially filled with water because ants hate water. And to further hinder the ants, the scientist painted the outside of the tub with a coat of very slow drying glue. He was sure that when he returned to his cottage the next weekend, the chocolate would be untouched.

When he returned, six days later, ants were swarming all over the chocolate. How did they do it? Well, the ants had crossed the glue because of the sacrifices of some of their heroes. It seems that a few ants had given their lives to form a single-file bridge of their bodies across the glue. Then they used pieces of grass and tiny pieces of wood to build a floating bridge to a leg of the stool, where it was an easy climb to the chocolate. The scientist also noticed that there were other daredevil ants crossing the ceiling, stopping over the saucer of chocolate, and then dropping right into their prize!

Proverbs 6:6 and 30:25 mention ants as examples of wisdom and industry, suggesting that those who are wise will learn from them. Wisdom is a gift of our Creator. Fear of the Lord is the beginning of wisdom, and Jesus Christ is the depth of His wisdom!

Prayer: Dear Father in heaven, the fact that all wisdom comes from You is a wonderful comfort in our world of foolishness and shallowness. Grant me true wisdom and understanding through Jesus Christ, Your Son, my Lord. Amen.

The Original Antibiotic

Genesis 1:28
"Then God blessed them, and God said to them, 'Be fruitful and multiply; fill the earth and subdue it; have dominion over the fish of the sea, over the birds of the air, and over every living thing that moves on the earth.'"

In Genesis 1:28 God tells man to subdue the Earth. That command implies that God has placed, within the creation, many tools that man can learn how to use for his benefit. After sin entered the world and sickness and death became a reality, God's remarkable foreknowledge became evident. One of the first antibiotics ever discovered by man has been in use for thousands of years. Modern researchers are just beginning to appreciate the wonder of this natural antibiotic which kills some 650 different strains of disease organisms, and is virtually non-toxic. Best of all, disease organisms don't become resistant to it.

What is this miracle antibiotic? Silver. The ancient Greeks and Romans used silver containers to keep liquids fresh. American settlers often placed a silver dollar in milk to delay it's souring. Most of the world's airlines today use silver filters on board to prevent dysentery. After testing 23 different methods for purifying water, NASA selected silver water filters for use on board the Space Shuttle. Japanese researchers have found that silver is even able to detoxify some poisons.

Isn't it striking that even the inanimate world is filled with things which are so carefully designed to fit into the overall picture of reality painted by Scripture. And who says the Bible isn't a book of science!

Prayer: Dear Father, we have only begun to scratch the surface in learning how to use all the wonderful things You placed into the creation for our benefit. We pray that You would guide us so that constructive discovery for the benefit of man may be done, and scientists would abandon destructive scientific work. In Jesus' Name. Amen.

Those Astonishing Bee Engineers

Proverbs 8:12a, 22, 30-31a
"'I, wisdom, dwell with prudence.... The Lord possessed me at the beginning of His way, before His works of old.... Then I was beside Him as a master craftsman; and I was daily His delight, rejoicing always before Him, rejoicing in His inhabited world....'"

The amazing structure of the honeycomb has fascinated scientists for thousands of years. In the third century, the astronomer and geometer Pappus of Alexandria became the first to offer an explanation for why the honeycomb has a hexagonal shape.

Pappus explained that only three shapes could serve as candidates for a honeycomb cell — the triangle, the square, and the hexagon. Any other shape would leave wasteful open spaces between each cell. Pappus noted that the hexagon holds more honey in the same space than either a square or a triangle. It also takes less wax to build, and the shared sides of the hexagonal cells cut wax usage even further.

But it was not until the development of modern calculus that scientists could fully appreciate the shape of the caps at the end of the honeycomb cells. Each cell is capped with a pyramid composed of three rhombuses. Complex mathematics shows that this shape too, requires the least amount of wax for construction and that it allows honeycomb cells to be butted up against each other without wasting space.

Modern scientists who accept evolution talk about the design of the honeycomb as a great accomplishment by bees. But the more sensible conclusion is obvious. The ten-sided prism of the honeycomb is magnificent testimony to the mathematical wisdom of the Creator Himself!

Prayer: Dear Lord Jesus Christ, I give thanks that I know You, the greatest of all wisdom, to be my Lord and my Savior. Teach me and use me so that those who are wise in the things of this world may be directed to true wisdom in Your saving work. Amen.

Spiderweb!

Proverbs 30:28

"The spider skillfully grasps with its hands, and it is in king's palaces."

Of about 35,000 known species of spiders, there are about 3,500 species which spin the familiar orb-shaped webs. These amazing, sticky silk constructions may be a few inches or even a yard across.

The first step in building an orb web is to cast out the highest horizontal silk line in the planned construction. Once both ends are anchored, the spider then pulls another strand below the anchor line into a "y" shape, then drops down while creating another silk strand to complete the leg of the "y". Other anchor lines may be attached from the center of the "y" to solid objects before the familiar spiral of the orb-web is started. Despite all of this work, most spiders take their webs down, eating the silk as they go, every day before the sun comes up. One question which nearly everyone asks, and scientists cannot even answer, is, why don't spiders get caught in their own sticky webs?

Scientists who believe in evolution have come to the conclusion that since so many different kinds of spiders build orb webs, the knowledge and ability to do so must have evolved many times. But we think it would be absolutely amazing if it even evolved only once!

Proverbs 30:28 refers to the spider's wisdom and skill, despite its small size. Other scriptural passages compare the most temporary of things in man's world to the spider's web. Scripture is true when it speaks of these earthly things. So why not accept the truth of Scripture when it speaks of other things that science studies, like creation itself?

Prayer: Dear heavenly Father, even the amazing spider shows forth Your wisdom and skill. Give me better sight so that I may more often see and appreciate Your wonderful works. In Jesus' Name. Amen.

Your Busy Liver

Hebrews 3:4

"For every house is built by someone, but He Who built all things is God."

An adult's liver is about the size of a football and weighs about three pounds, making it the body's largest internal organ. Tucked neatly beneath the ribs, your liver performs more than 500 different tasks. It is a vital link between your heart, lungs and digestive system.

Inside the liver is a bewildering array of microscopic veins in which each drop of blood is processed. Here, blood conditions are constantly monitored to make sure everything is up to standard. If more of certain substances are needed in the blood, they are supplied. Useless chemicals are broken down into useful chemicals. Proteins are made in the liver, blood clotting factors are corrected, hormone balances are maintained, and poisons are neutralized. If substances are needed to fight an infection, they are created and added to the blood.

The liver also stores vitamins and minerals and prepares itself to provide your body with quick energy when you need it. In addition, the liver makes bile, which is essential for digestion.

Structures like the liver have caused many evolutionists to abandon the idea that life is a result of millions of years of accidents. The liver is just too well designed and integrated into the body to have been produced by purposelessness and mindlessness.

Prayer: I thank You, Dear Father in heaven, for the wonderful way in which You have made me and for the ways in which You keep my life going. Help to dedicate my life to You, not only in word, but also in deed. In Jesus' Name. Amen.

How to Speak Bee

Psalm 26:6-7
"I will wash my hands in innocence; so I will go about Your altar, O Lord, that I may proclaim with the voice of thanksgiving, and tell of all Your wondrous works."

Which animal has the most sophisticated, specific, and abstract language? Since evolutionists are always linking language with intelligence, and man with animals, one would expect that perhaps the ape or some similar primate would turn out to be the best animal communicator. But evolution's expectations about language, and a link between man and apes, proves to be wrong.

The bee has the most complex and detailed functional language among the animals. Using a formal coded dance, sounds, and smells, a bee can communicate an extraordinary number of details to its hive-mates. Let's say that a bee has located a particularly good source of nectar and returns to the hive to tell its friends what direction and how far away it is. Its dance communication would form a figure eight, with the cross points at the center of the eight giving the direction of the nectar in relation to the sun. As the bee dances on the wall of the honeycomb, the position of the sun is always down. If the bee moves up the comb wall at 17 degrees to the left of vertical, it means the honey is 17 degrees to the left of the sun. This communication even works on cloudy days, since bees can see the ultraviolet light which penetrates the clouds. Distance is communicated by the speed of the dance.

Neither man's languages, nor the bees' language evolved — they were given by God. Why else would the bee measure degrees of inclination of the sun in the same way that human beings do? That the bee is the best communicator in the animal world, and not some ape-like creature, helps to show that language cannot be explained by evolution.

Prayer: Dear Father in heaven, I am filled with wonder and rejoicing when I hear about all of the wonderful things You have created and how You have provided for all of Your creatures. But never let me forget to rejoice in Your salvation. In Jesus' Name. Amen.

Ant Mathematics

Luke 14:28-30

"'For which of you, intending to build a tower, does not sit down first and count the cost, whether he has enough to finish it — lest, after he has laid the foundation, and is not able to finish it, all who see it begin to mock him, saying, "This man began to build and was not able to finish."'"

Can ants count? It seems so! When a scout ant find an item of food, they take it back to their nest. But if the food item is too big to carry, but especially good, the scout will return to the nest to get help. Scientists have discovered that ants apparently size up the task ahead before getting help so that they can return with enough, but not too much, help.

One scientist cut a dead grasshopper into three pieces. The second piece was twice as big as the first, and the third was twice as big as the second. He then left the pieces in different locations where ants were sure to find them. He watched as each was discovered by a scout, inspected, and each scout returned to the nest for help. When the scout returned with help, the scientist counted the number of ants working at each piece of grasshopper. The smallest piece had 28 ants working on it. The piece which was twice its size had 44 ants working on it, almost twice as many as the smallest piece. And how many ants do you think were working on the piece that was twice the size of the second? If you doubled that 44 to 88 you would be within one of being right — there were 89 ants working to return the biggest piece to the nest!

We cannot help but conclude that mathematical ability is part of the ants' amazing ability to plan and carry out a task! They were doing nothing more than their Creator taught them as they followed the planning principle that Jesus reminds us of in Luke 14:28 — count the cost of the project before you begin it!

Prayer: Dear Lord Jesus Christ, You spared nothing for my salvation, even though it cost You terrible suffering and death. As Yours, give me wisdom to plan well in the things of the world, but no desire to withhold anything from You because of the cost. Amen.

The Puzzle of Mathematics

Genesis 1:31a

"Then God saw everything He had made, and indeed it was very good."

One of the best ways to find out whether a piece of clothing is of high quality is to look at the quality of the fabric which has been used to make it. Usually, someone who selects a high quality fabric is concerned enough with quality to put good workmanship into their product. And likewise, the fabric of the creation tells us a lot about its Creator. Mathematics more than anything else, reflects the fabric of the creation.

Many modern scientists are puzzled by the fact that the material world they study can be accurately described in mathematical equations. The falling of a stone or a feather, even the colors of a rainbow, all correspond to mathematical formulae. Even chaos has a mathematical description. One Nobel physicist was moved to write a paper entitled, "The Unreasonable Effectiveness of Mathematics in the Natural Sciences."

Because mathematics so precisely describes reality, we can send a space probe billions of miles, on perfect target. Mathematics allow us to receive the pictures it returns to Earth using a transmitter no more powerful than a five-watt light bulb. As one science writer put it, scientists use mathematics as a wonderful gift but they have no idea why it works so well.

Of course, creation-scientists don't wonder why math works. They know that the creation is the work of an orderly and wise Creator Who was pleased with His final product. The precision of mathematics reflects the excellent quality of His fine workmanship.

Prayer: Dear Father, I praise You because You have truly done all things well. Help me to be a better witness to Your excellent workmanship to those around me. In Jesus' Name. Amen.

Bees Outsmart Scientists

Psalm 19: 9, 10b
"The fear of the Lord is clean, enduring forever; the judgments of the Lord are true and righteous altogether . . . sweeter also than honey and the honeycomb."

Scientists have long recognized that bees are complex and intelligent creatures. Unfortunately, many scientists continue to think of bees as the product of millions of years of chance evolution, despite the fact that new research reveals the fingerprints of God more clearly than ever.

Recently, researchers from Princeton University decided to find out whether bees were smart enough to find their food source if it was moved. Researchers moved the prime food source for one hive 50 meters farther away from the hive. They found that it took the bees less than one minute to locate the moved source. Being precise scientists, they then moved the food source fifty additional meters away. The bees still took less than a minute to find the food source. Two more moves, each a precise 50 meters, produced the same results.

However, the bees had also been studying the researchers. Before the researchers could finish moving the food source yet another fifty meters, they found that the bees had discovered the pattern and were already waiting at the new location!

God makes the fact that He is our Creator evident in the creation so that we will seek the saving details about Him and His love for us in Jesus Christ in the pages of the Bible. While the earthly honey made by bees is sweet and delightful, it does not compare with truly knowing Him as He shows Himself to us in Scripture. Seek Him there.

Prayer: Dear Father in heaven, help me to have eyes which not only see Your creation so clearly that I can discern Your fingerprints, but so that I can also clearly see what You are teaching me in Your Holy Word. In Jesus' Name. Amen.

Fast Rocks

Jeremiah 2:26-27
"'As the thief is ashamed when he is found out, so is the house of Israel ashamed; they and their kings and their princes, and their priests and their prophets, saying to a tree, "You are my father," and to a stone, "You gave birth to me," ... But in time of their trouble they will say, "Arise and save us."'"

Unfortunately, many people reject the Bible's claim that it reports actual history because they imagine that there are other ways to measure the Bible's truth. The Bible makes very clear claims about the origin of the world.

Because rocks seem so solid and permanent by human standards, it is easy for many people to believe that rocks last virtually forever. It is also easy to believe evolutionary claims that rock formations represent millions or billions of years of Earth history. Every year, millions of visitors to natural caves are treated to claims that the stalactites and stalagmites growing within the cave represent hundreds of thousands to even millions of years of natural growth. Like many other evolutionary claims, this one is easy to disprove. Stalactites are said by evolutionists to require a century to grow one inch. Yet foot-long stalactites have been found in coal mines in the lime-rich area of Newcastle, Australia. The mines in which these are found are less than 40 years old. Similar examples can be found in the Sequoyah caves in the United States.

In His Word, our Creator warns us many times about looking to the creation as our source of life and living. Unfortunately, when we think that the creation is able to correct His Word, we are guilty of the same idolatry that atheistic evolutionists commit. Yet our gracious Creator continually reminds us He is there, waiting to forgive us, for the sake of the saving work of His Son Jesus Christ on our behalf.

Prayer: Dear Lord Jesus Christ, You were the Instrument of our creation and You are the Word made flesh. Cleanse our hearts of all thinking which would replace You with what You have made and grant us Your Spirit to make a strong witness for Your Word. Amen.

Non-Reproducible Reproduction

Genesis 1:31
"Then God saw everything that He had made, and indeed it was very good. So the evening and the morning were the sixth day."

There are many examples of ways in which God's creative mind came up with interesting and unexpected surprises when He designed the creation. One such surprise is the way a species of common thrip gives birth. The thrip is a gnat-like bug which lives in and helps break down decaying leaves. Yet this common and seemingly unimportant insect has been given an ability by its Creator which has been seen in no other creature!

Many of us can remember how, in biology class, we had to learn how creatures reproduce. Some give birth to live young while others lay eggs. But this common species of thrip, science has learned, does both! And that is not the end of the surprises. The thrip reproduces twice a year. At one time of the year the female thrip lays eggs. All the thrips which hatch from these eggs are females. But during the other mating period, the female thrip produces only male offspring. The male thrips are live-born and ready to begin the typical life of a thrip as soon as they are born. Researchers continue to study the thrip to find out how this strange arrangement can benefit the thrip.

It is interesting that researchers think that there is some benefit to the thrip in this arrangement. Yet science usually suspects that there are benefits to unusual arrangements in nature because they usually find that such benefits exist. And all by itself this is a testimony to the Creator Who has made His individual care for each creature so easy to see that even evolutionists expect to find wisdom in creation's design.

Prayer: Dear heavenly Father, I often forget just how creative You are, especially when my problems seem so great and no solution seems possible. Help me to have a strong faith which remembers how creative You have been in making the tiny thrip, and to realize how much more You love me. For Jesus' sake. Amen.

The Adder Actor

Matthew 10:16
"Behold, I send you out as sheep in the midst of wolves. Therefore be wise as serpents and harmless as doves."

Some animals use quite a bag of tricks to make their livings or protect themselves. Who taught them these things? If you believe what the Bible says about creation, the intelligence — even wisdom — found in the animal world is easy to explain.

The hognosed snake or puff adder is one example of a wise serpent. The puff adder is a harmless fellow who doesn't like to be bothered, especially by people. You might find one sunning itself as you walk across a field. If he sees that you see him, he will coil himself up so that if you try to grab him, you will have to reach past his head. Then, to make you think twice about reaching past his head, he puffs his head up just like a cobra. No one wants to grab a cobra! If you just sit and wait a bit, the snake will realize that his plan isn't working, so he'll go to plan "B" — he'll strike at your leg, pretending to bite. He won't bite, but he hopes you'll run off. If this doesn't send you running, the puff adder goes to plan "C" — writhing on the ground as if you are beating him and he is in terrible pain. Finally, he turns over on his back, tongue hanging out of his mouth as if he were dead. The puff adder continues to play dead even if you pick him up and handle him. He'll stay "dead" unless you put him back onto the ground on his stomach. If you do that, he'll flip over and return to his original "dead" position.

Our Lord warned Christians not to be foolish, but rather to be wise like the serpent, yet harmless as doves. The wisdom He has given to the animal kingdom is nothing compared to the wisdom He will give Christians who will allow themselves to become His instruments.

Prayer: Dear Father, You have taught the creatures of Your creation so many things that even we can learn from them. Do not let me be satisfied with this wisdom alone, but rather seek Your deeper wisdom in the pages of Your Word. In Jesus' Name. Amen.

Space-Age Vindication for the Bible

Job 26:7

"He stretches out the north over empty space; He hangs the Earth on nothing."

The Earth floats in space, attached to nothing, surrounded by a thin layer of air. What science has only just learned, the Bible has taught for thousands of years! While other ancients pictured the world as flat, or resting upon giant turtles or some other animal, God told the Jews, in Job 26:7, that He hangs the Earth on nothing.

In Genesis 1:6 and 7 we read that God created a firmament. In recent times, some have said that this description of the firmament proves that the Bible is based on ancient myths. New discoveries, however, are challenging these doubts about the Bible. The word translated "firmament" in these verses comes from a Hebrew root word which refers to the process of making a statue. In making a statue, the ancient artisan would take gold, a very soft and easily worked metal, and begin to pound thin sheets of it onto a wooden form of the statue until the wood was completely covered by a thin, form-fitted layer of gold.

The use of the word translated firmament puzzled many people until recent times, when the Earth was first viewed from space. Then we saw it — the Earth, literally hanging on nothing in space, surrounded by a thin, form-fitted layer — our atmosphere!

The Bible tells the truth in all the subjects it mentions. But no matter how long science studies, it cannot learn about God's love to us in Jesus Christ. This is revealed to us only by the Bible!

Prayer: Lord, Your Word is truth. Strengthen us in Your truth so that we may not be misled by the false claims which label Your revelation to us "out of date." Help my faith to be more than just good intentions; help my faith to be well-informed by Your Word of Life. In Jesus' Name. Amen.

Yeast Enters Electronics Production

Exodus 12:15
"'Seven days you shall eat unleavened bread. On the first day you shall remove leaven from your houses. For whoever eats leavened bread from the first day until the seventh day, that person shall be cut off from Israel.'"

Probably the first thing that comes to your mind when yeast is mentioned is its use in bread making. Even 3,400 years ago, when Israel left Egypt, yeast was common in making bread.

Now physicists at AT&T Bell Laboratories in Murray Hill, N.J., have learned that yeast has yet another amazing ability. For years scientists have been trying to make semiconducting crystals for modern electronics which are so tiny that they are like individual molecules and atoms. Such tiny electronic parts would result in another giant leap in modern electronics. But scientists have had difficulty making such crystals consistently small enough. Now we know that yeast can do it. When fed cadmium sulfide, yeasts try to detoxify the metal, and in the process, create exactly the tiny electronic crystals scientists have been looking for. It is very possible, as a result of this discovery, that you may someday bake bread which rises because of yeast in a microwave oven that is controlled by a computer containing tiny electronic parts also made by yeast!

Because the Bible is a real-life book which tells about a real-life God — the true God — we should not be surprised to learn that this Creator is interested in man's real-life, everyday activities. One result is that we can explore what He has made, always expecting to learn some new thing which He has placed in the creation to help us. A more important result is that we see how excellent and true the real life, saving Word of our Creator is for us today!

Prayer: Dear Lord Jesus Christ, I thank You that I can clearly see that You are interested in all aspects of my life. Help me to remember Your all-encompassing love and care as a I face life and as I help others face life with You. Amen.

Solitary Bees

Proverbs 17:17
"A friend loves at all times, and a brother is born of adversity...."

When most people think of bees, they think of highly social insects which live together in large colonies or hives. In fact, not all bees live in hives. Bees, like people, can actually be found in a variety of social relationships, including those that hardly have any other contact with others.

Like people, most creatures need friends. Some female bees of a species which lives in the ground make it a point to build their homes next to another female. They often connect their separate underground homes with a tunnel so they can visit each other like best friends. Sometimes they even lay their eggs near each other and raise their young together. Often one female will baby sit both sets of young while the other goes out for groceries.

The almost universal need for companionship among living things should tell us something about the One Who created all of them. Our need for companionship tells us how important it is to have others with which we can share life. And it should tell us that this is something that our Creator values too. In fact, the message that is part of every page of the Bible is that He made mankind so He would have someone to love.

Is He your Number One companion in your life? Do you know His love for you through His Son, Jesus Christ? Whether you don't know Him at all, or whether you would like to know Him better, He comes to speak His Word to you in the pages of the Bible.

Prayer: Dear Father, although You have made me for the purpose of having a loving relationship with You, I confess that at times I have withdrawn my love from You. Forgive me for Jesus' sake, and help me to make You my Number One companion. Amen.

Wasps Command Aphids

Matthew 13:39
"'The enemy who sowed them is the devil, the harvest is the end of the age, and the reapers are the angels.'"

The ability to control someone else's will is the stuff of classic horror movies. Science fiction movies aside, man has thankfully never learned how to take over another person's will to the point of controlling their every action. But a recent scientific report says that a certain parasitic wasp may, in fact, be able to control the actions of the aphid which it attacks.

This tiny, parasitic wasp injects its eggs into the body of aphids. As the eggs develop, the larval wasps are nourished by the aphid's body. Once the aphid dies, the wasp pupae incubate until they emerge as adult wasps. It's no surprise to learn that aphids object to this. Some aphids will actually commit suicide by jumping to their deaths after they have been injected with eggs. Others, according to scientists, seem to follow the commands of the tiny wasp parasites within them. Aphids carrying wasp eggs which will require hibernation over winter often leave their plants to die in a protected spot. There the young wasps will have the best chance of making it through the winter. Aphids carrying eggs which will hatch before winter remain on their plants to die to provide the best place for the young wasps to start their adult life.

As evil as all of this sounds, we human beings face a more dangerous threat than these aphids. Jesus Christ pointed out that until He returns to this world at the end of time, His enemy and ours, the devil, will try to get us to follow his commands. But unlike the aphids, we have a way out of his evil control because of what Jesus Christ Himself has done for our salvation.

Prayer: Dear Savior, I thank You that You have not left me without hope. I ask that, no matter how attractive the devil's lures appear, each of his attacks may drive me closer to you. Amen.

God's Joy of Creating

Romans 11:33
"Oh, the depth of the riches both of the wisdom and knowledge of God! How unsearchable are His judgments and His ways past finding out!"

Have you ever tried to plan all of the details of a simple project? How many plans do you think the Lord had to make when He created living things? A billion? A billion times a billion?

We all know that it takes time to plan the most simple project. But did you ever think about the planning God had to do when He created all the different kinds of living things? Our word "species" includes many creatures which the Bible counts as being the same "kind" — as when God created the kinds. But God designed the genetic information which allowed the kinds to produce many variations. God's act of creating living things was much more than simply wishing. Just think, there are more than 20,000 different species of bees — some with very complex societies and their own languages! There are over 7,000 kinds of segmented worms.

The numbers and the beauty of it all makes one wonder at God. Why are there 4,500 different species of sponges? Why are some creatures, never seen by man until this century, so eerily beautiful? For that matter, why do we have so many different kinds of beautiful flowers?

The variety in the creation reflects some of the joy of creating that God felt as He made the world. This variety shows us the incredible unbridled creativity of our wonderful God. And amidst all this, the fact that there is only one species of man — all related — helps confirm the history of man related in the Bible.

Prayer: Dear Heavenly Father, I look forward to the day when I shall see You face to face and my understanding shall grow in Your presence. But do not let me be satisfied to inactively wait until that time comes to praise You before men. In Jesus' Name. Amen.

Too Much Knowledge?

Job 37:16
"Do you know the balance of the clouds, those wondrous works of Him Who is perfect in knowledge?"

Most of us are familiar with the fact that our Earth is being observed by dozens of satellites which keep track of weather systems, natural resources, ocean currents, sunlight, climate, and even animal migrations. As scientists are learning more about the many factors which make life on Earth what it is, they are also beginning to discover how much they *don't* know.

For this reason, scientists plan to have a new system of observational satellites in operation by the mid-1990's. This system, called EOS, for Earth Observing System, will include at least four satellites which are able to monitor just about any weather, climate, or atmospheric activity taking place on Earth. It is expected that these satellites will collect about 1 trillion bits of information per day. In less than its first week of operation, the amount of data gathered by the system will be greater than all the geoscience information recorded in man's history. Scientists are beginning to wonder how they are going to organize all that information in a usable way.

More than ever, the EOS project reminds us that without a Creator to put the information into the Earth's weather and geological systems, there would be no geoscientific information for scientists to gather. And despite the huge amount of information which EOS will gather, it is only a small portion of all that is going on with the Earth. Although our information is still so limited that man worries about how he will handle it, it is only a small amount of all the information God has created.

Prayer: Dear Father in in heaven, I am awed by the size and scope of the creation You have made. But I am even more in awe of Your love which led You to send Your Only Son for my salvation. In His Name. Amen.

Aphid's Clever Defenses

Job 39:26
"'Does the hawk fly by Your wisdom, and spread its wings toward the south?'"

Unless you have roses, you probably don't care one way or the other about aphids. Often called plant lice, aphids are tiny creatures which are often tended — even protected — by ants. Like many creatures, aphids provide wonderful testimony to God's ability to create endless variations on a theme.

Aphids come in many different colors: white, red, green, and yellow. Aphids look very defenseless. But instead of relying on teeth or stingers, aphids rely on cleverness to protect themselves. Some ants also protect the aphids they tend. But aphids have two spikes on their backs. These spikes are not for stinging, but for gluing. Lady bugs love to eat aphids. If a lady bug tries to eat the aphid, the aphid tries to position its spikes to glue the mouth of the hungry lady bug closed. Aphids which are not protected by ants have longer, more effective spikes.

Perhaps the most clever strategy aphids use to protect themselves is the "aphid kick." When a group of aphids sitting on a plant feel threatened, they will kick their hind legs in unison. This seems to make the would-be predator think that something bigger is after him!

Nothing God has made is unimportant. If nothing else, the simple fact that the Creator wants something to exist makes that thing important. After all, it's important to Him! You can be thankful that no matter what other people may think of you, God has made you — and that makes you important to Him. But He has done more to assure you of His love for you by giving you His Word in the Bible.

Prayer: Dear Heavenly Father, I thank You that You have made me. Do not allow me to separate myself from You by neglecting to read and study Your Word of Life, wherein I learn of my salvation. In Jesus' Name. Amen.

Bee Environmental Engineers

Genesis 1:25
"And God made the beast of the Earth according to its kind, cattle according to its kind, and everything that creeps on the Earth according to its kind. And God saw that it was good."

Man has always been interested in living in comfortable temperatures. Excavations from the most ancient human sites show that homes and entire cities were often positioned to catch the prevailing winds in the summer or to avoid cold blasts in the winter. Obviously heated homes go back much further in man's history than do air conditioners. However, inventors have been devising air conditioning schemes for buildings for hundreds of years.

Honeybees have been successfully keeping their hives precisely heated or cooled for thousands of years. Honeybees prefer to keep their hives at a constant 95 degrees F and their way of doing it is quite ingenious. When the weather is cold bees collect at the center of the hive where they generate extra heat by increasing their metabolisms through rapid breathing. Other bees position themselves around the sides of the hive to serve as insulation. If it remains cool outside, the bees at the warm center of the hive rotate with the bees at the cooler walls of the hive. If the weather gets too warm some of the bees act as fans, circulating outside air into the hive. If this does not provide enough cooling, other bees leave the hive to bring water back which will be spread on the hive walls by other bees. Now the fanning of the other bees will actually cause the walls of the hive to cool.

The intelligence we see at every level of the creation offers strong evidence *against* the claim that intelligence is a product of evolution. Intelligence is a gift of the Creator to His creatures!

Prayer: Dear Lord Jesus Christ, I thank You that You have made me with the intelligence to appreciate at least some of what You have made in the creation. Help me to make better use of my intelligence to Your glory. Amen.

Faith and Sense in the Origins Debate

Hebrews 11:1
"Now faith is the substance of things hoped for, the evidence of things not seen."

As Christians, we walk by faith. But did you know that even evolutionary scientists walk by their own kind of faith? Early in this century scientists had no evidence that man had evolved from ape-like creatures — just as they don't today.

In 1922, a tooth was discovered in Nebraska which was said to have belonged to a missing link between man and ape. What did this creature look like? As is done today, paleontologists began to "reconstruct" "Nebraska man." They "reconstructed" what the jaw bone around the tooth might have looked like, and then the bone touching those bones, and — well you know how the song goes: "The head bone's connected to the neck bone; the neck bone's connected to the back bone . . . ." Before long, they had "reconstructed" from that one little tooth, not only what "Nebraska man" looked like, but also what his wife and child looked like. And they put this "proof" in museums and textbooks. Eventually, they discovered more parts of the animal from which the tooth had come. It turned out to be the tooth of an extinct pig!

Sure, anyone, including scientists can make mistakes. But what this true story shows is how, without any evidence and only a faith which rejected God's account of man's creation, a pig could be made into a man. You see, even the evolutionist walks not by scientific fact, but by his own faith in the creation instead of the Creator. We Christians should not be ashamed to admit that we walk by faith, because our faith is built on the solid statements of the uncontradicted Scriptures!

Prayer: Dear Father, I thank You that You have allowed me to hear Your Word, and that You have given me faith in Your promises. Teach me, through Your Word, so that I am better able to identify false religious beliefs and carry Your witness to others. In Jesus' Name. Amen.

Male and Female: More than Genetics!

Genesis 1:27
"So God created man in His image; in the image of God He created him; male and female He created them."

The currently popular idea that the only differences between men and women are unimportant and incidental grows out of the evolutionary view of the world. According to evolutionary theorists, male and female is only a biological experiment — and some say it is not a very good one! The Bible takes a very contrary view, extolling the unique virtues and abilities of both men and women. The biological sciences are beginning to see the wisdom of this design.

You remember learning in biology that each of us started in life when genes from our father were united with an equal number of genes from our mother producing a fertilized egg. Perhaps you have even heard the flippant evolutionary claim that all it takes is two sets of human genes to make a human being. But in 1984, researchers tried to fertilize mouse eggs with equal sets of mouse genes from other females. They learned what Scripture has said all along — there's more to it than simple chemical mechanics. For their trouble they got no mice. Scientists have learned that there are very real differences between identical chemical structures produced by males and females. They have also discovered one of the purposes this difference serves; the male proteins on the surface of the developing fetus and placenta modify the mother's immune response so that she does not reject the growing child.

God's wisdom in creating male and female, with their very important differences, cannot be denied — and it's being supported by modern science!

Prayer: Dear Lord Jesus Christ, I thank You that we have been made male and female. If it had been up to us to make ourselves, we obviously could not have done it. Help me to be able to witness Your truth to those who believe that the creation made itself. Amen.

Has the Big Bang Gone Bust?

Psalm 136:4-5

"... To Him alone Who does great wonders, for His mercy endures forever; to Him Who by wisdom made the heavens, for His mercy endures forever...."

Most of us are familiar with the Big Bang theory. It says the entire universe and everything in it is the result of a gigantic explosion which took place about 16 billion years ago. This explosion eventually resulted in stars, galaxies, our Earth with its delicate balances, and even our genetic codes. The problem is, we are all familiar with explosions and no one has ever seen an explosion make anything more than destruction. Creation-scientists have been saying for years that an explosion could not have created the universe we see today.

The news is that evolution-scientists are beginning to agree. Most of them still think that an explosion could create the Earth we know today, but they are finding that the organized way in which stars and galaxies are scattered through space is exactly opposite of what their theory leads them to expect. One scientist has suggested that maybe they are not taking everything into account. He added that they are looking for ideas — to use his words, even "crazy" ideas, "because, we're getting a little desperate."

Of course there is a solution to their problem. And the solution isn't so "crazy," either. You don't get words in a book without an author. You don't build a house without a plan. And you can't get a universe like ours which is literally crammed with information and information storage systems without a Creator.

Prayer: Dear Heavenly Father, as I gaze into the star-filled night sky, or at a beautiful flower, or ponder the wonder of my own body, I cannot help but see that You, and no explosion, have created me and all things. In Jesus' Name. Amen.

The Insect Without Children

Jeremiah 51:15
"He has made the Earth by His power; He has established the world by His wisdom, and stretched out the heaven by His understanding."

Did you know that there is an animal that can have a huge number of offspring, yet never have any children? If you think that sounds mysterious, you'll be happy to know that science has no idea how the system works.

That unusual creature is the aphid. The first aphid of spring is a female, has no wings, and hatches from an egg very early in the season. After about ten days of feeding, she will give live birth to another female aphid. There have been no males around, and her offspring is a fully-formed adult. The first thing it does is feed. Within ten days the new female, too, is ready to give live birth to another wingless adult aphid. This can go on all summer — 2, 4, 16, and so on. By the end of the summer, that first female aphid could have 5,904,900,000 offspring! Obviously one plant can't hold that many aphids. So when a plant has all the aphids it can support, the offspring are born with wings so that they can fly off to another plant to continue this mysterious process. Once cold weather begins, the female aphids begin giving birth to both male and female aphids, some with wings and some without. The winged offspring fly back to the original plant, mate, and lay eggs that can withstand winter so the entire cycle can start again the next year.

The word "understanding" is often used in Scripture to describe how God created. In this sense, "understanding" refers to the sense in which God created all things in such a way that they interact and react to countless situations because he predesigned them that way. The lowly aphid is an astonishing example of this!

Prayer: Dear Lord Jesus Christ, I am filled with awe and wonder when I think that You, the Instrument of creation, came to our Earth and took on our form to save us from the consequences of our sin. Help me to aspire to the same kind of love and wisdom. Amen.

Earthworms

Psalm 104:30
"You send forth Your Spirit, they are created; and You renew the face of the Earth."

There are many things in creation which people seldom talk about. It's almost as if some things are not worth our attention. But since God made them, they were worth His attention. And since God is infinitely greater than we are, maybe that should teach us a lesson.

Earthworms are considered to be among the lowliest creatures on Earth. Yet they are important for our soil, preventing it from compacting, helping water reach plant roots, and helping vegetable matter decay and make the soil rich. In addition, earthworms increase those good bacteria that further enrich the soil for plant growth. That makes earthworms important.

When He made earthworms, God could not resist making a great variety. There are over 170 kinds of earthworms. You have probably seen everyday earthworms and night crawlers. But some earthworms in Australia grow to a length of 10 to 12 feet! Each worm has a well-developed nervous system — not simple as evolutionary theory expects. The earthworm also has five pairs of pumps which serve as hearts. Each worm has microscopic bristles which help it crawl through the soil.

Psalm 104 talks about God's faithfulness. Verse 30 mentions how God renews the creation with new spiritual life through His Holy Spirit, and in the same breath speaks of how He renews the soil to support material life. Earthworms play a major part in God's renewing of the soil. How wonderful to see God's Hand everywhere we look!

Prayer: Dear Lord Jesus, help me to have clear vision so that I can see all things by faith, discerning God's agents and acts, no matter how lowly they appear to earthly eyes. Amen.

A Solid Foundation in Creation

John 1:3
"All things were made through Him, and without Him nothing was made that was made."

What does the Bible's teaching on creation have to do with the gospel of Jesus Christ? Are these just two separate and unrelated biblical teachings? Christians often ask us, "Shouldn't we just worry about the basics? Why worry about creation?" Indeed we should be concerned with basics.

But what happens to Christ's work of salvation if creation is not true? If evolution is true, then death came into the world long before the first man and before sin. If death is not a result of sin, why did Christ have to receive the penalty for sin, death on a cross? So, in challenging man's origin, evolution challenges the origin of sin and its effect on man. Thus, evolution challenges the very heart of Christ's work! The Bible tells us that all Scripture was given to make us wise unto salvation. That includes Genesis. The first chapter of St. John's Gospel tells us that all things were created through the Word — the Word Who became flesh.

We see that Word in action in Genesis 1 when we read, "And God said . . . ." That Word Who made us and everything is the very same Word Who came and purchased our salvation.

So Genesis, beginning with the first chapter, is actually the beginning of God's revelation to us of the Person and work of the Son of God — our Savior Jesus Christ. If we reject God's revelation about our Savior in Genesis 1, we have only part of a Savior — and only part of Christ is no Christ at all!

Prayer: Dear Lord Jesus Christ, many in our world today attack You by denying Your work of creation, hoping to deny Your wonderful saving work for sinful mankind. This even happens in the church. Give Your people eyes to see the truth and not be led into error and left with a powerless word of man in place of the gospel. Amen.

Coming: A Robot Dancing Bee

Zechariah 7:9-11
"Thus says the Lord of hosts: 'Execute justice, show mercy and compassion everyone to his brother. Do not oppress the widow or the fatherless, the alien or the poor. Let none of you plan evil in his heart against his brother.' But they refused to heed, shrugged their shoulders, and stopped their ears so that they could not hear."

In past *Creation Moments*, we have talked about how honeybees communicate with each other through dances. These dances are used by scouts who have located a source of food to communicate the location, direction, and distance of the food source to other bees. Researchers have now learned that there is more to this communication than simply dance.

Scientists have proven that bees can also hear the buzzing sounds which accompany the dance. Further, they have been able to mimic the buzzing sounds, but they have not been able to reproduce the language which bees use. They suspect that bees hear sounds using organs which are located at the base of each antenna. In order to further to study bee communication, researchers in Denmark plan to construct a tiny dancing bee robot which can be used to recruit real honeybees to food sources — if they can learn the bees' language.

In a previous *Creation Moments*, we also talked about the recent discovery of an ancient bee encased in amber. Scientists said that even though the bee dated back to the beginning of flowering plants, it was just like modern bees. The Lord's message through all these lines of evidence — sophisticated bee society and communication and the evidence that it has existed since there were bees — is clear. If dancing bee robots must be manufactured by intelligence, the real dancing bees must have been made by an intelligent Creator too.

Prayer: Dear heavenly Father, it is sometimes perplexing to see how deaf the world is to what You say so clearly. Yet I confess that I, too, don't always hear Your Voice when I should. Help me to hear You better and to lead others to hear Your call to them. In Jesus' Name. Amen.

Another Evolutionary Myth About Man

Genesis 4:2b
"Now Abel was a keeper of sheep, but Cain was a tiller of the ground."

According to man's history as presented in the Bible, farming and shepherding were among the first occupations. While Adam may have followed one of these occupations, Genesis specifically tells us that there was a shepherd, Abel, and a farmer, Cain, among the second generation of human beings to ever live. Evolutionists say that this history is wrong — the first men were hunters and gatherers.

In order to support this view, scientists are quick to announce the discovery of primitive tribes which they think are not advanced enough to farm or keep their own animals. When they were first discovered in 1962, the Agta people of Luzon were introduced to the world as "living fossils" of what all of mankind once was. Later, one researcher who lived with the Agta for some time so that he could study them, heard someone singing in English outside his tent one night. One of the Agta women was singing her baby to sleep with a religious song in English. So much for the innocent, primitive, "living fossil" theory! Similar long-term studies of other hunter-gatherer groups around the world have shown that the whole idea of "living fossil" hunter-gatherers is a myth. A 1989 issue of *Science News* carried an article on the subject with the interesting title of, "A World That Never Existed."

Man's attempts to explain himself come and go, but what the Word of God says about man will stand forever. In fact, our eternal life itself is revealed in the Word of God!

Prayer: Dear Lord Jesus, You took our form upon Yourself, a form which You created, to win our salvation. Help me to be better able to show those around me how excellent the Bible's message about man truly is. Amen.

The Woodpecker's Amazing Tongue

Deuteronomy 4:28

"'And there you will serve gods, the work of men's hands, wood and stone, which neither see nor hear nor eat nor smell.'"

The woodpecker's tongue can stretch from three to five times its normal length in order to fish bugs out of trees. You would think that its tongue would have to be rooted in its tail to do that neat trick! The beautiful creation in which we live is not only filled with testimonies to the wisdom of God, it is also filled with special designs which deny the possibility that the lifeless creation made itself and all life.

Consider the woodpecker. Its tongue is rooted in its right nostril. Exiting the nostril, the tongue splits into two parts, wraps around its head between its skull and the skin, passing on either side of the neck bones and then coming up through its lower jaw, or beak. This gives the woodpecker a long enough tongue to stretch it far enough to do an effective job of pest control on bug infested trees! Now how could that happen by blind evolutionary chance?

Even evolutionists admit that it's silly to suggest that the woodpecker's tongue, gradually over thousands of years, got longer and began to grow under his skin. As one evolutionary scientist said about the woodpecker's tongue, "There are certain anatomical features which just cannot be explained by gradual mutations over millions of years. Just between you and me, I have to get God into the act too sometimes."

Why wait to call on God as a last resort? Let's *begin* with our wonderful Creator!

Prayer: Dear heavenly Father, through the instruction of Your Word and the guidance of Your Spirit, help me to be different from those around me who think that the creation itself made them. In Jesus' Name. Amen.

The Cap-Throwing Fungus

Genesis 1:11
"Then God said, 'Let the earth bring forth grass, the herb that yields seed, and the fruit tree that yields fruit according to its kind, whose seed is in itself, on the earth,' and it was so."

God doesn't make cheap things! While we speak of "simple" forms of life, the more we learn about living things, the more clearly we see that there are no simple forms of life.

Evolution has long theorized that life began with simple forms which gradually, over millions of years, became more complex. Evolutionary scientists say that the "simple" forms of life still in existence today prove their claim. The problem is, there aren't any simple forms of life! Take the "lowly" fungus, for example. The "cap-throwing" fungus has a number of designs which enable it to spread its reproductive spores. The "cap-throwing" fungus has a built-in clock and bends through the day in response to the sun's movement. It also throws its masses of spores out to spread them over the widest possible area.

The built-in clock of the "cap-throwing" fungus waits to "blow its top" until the fungus is turned at just the best angle to produce the widest possible spread of its spores. The light-sensing system in the fungus releases the spores at about nine in the morning — aiming the spores at an area that is most likely to be open so that they can be spread even farther by animals. Whether the spores land where animals pass, or on a leaf, they are coated with a glue to aid in further dispersion.

The lowly fungus reminds us that there just aren't any simple forms of life. This fungus has been given sophisticated ways to carry out its command to reproduce after its kind.

Prayer: Dear Father, there are none of us who are so simple or unimportant that You have not enabled us to do Your will. Help me to always remember that the hindrance of sin is of my making, and to come to my Savior Jesus Christ for cleansing. In His Name. Amen.

Moths that Think They're Hummingbirds

Psalm 92:5-6
"O LORD, how great are Your works! Your thoughts are very deep. A senseless man does not know, nor does a fool understand this."

There seems to be no end to the variety and wisdom of design in the creation. Most interesting are those creatures which share important traits and yet are nothing like each other. While that may seem to be a contradiction, consider the whale, which lives and is constructed much like a fish, but otherwise shares the characteristics of mammals. The sphinx moth is another such creature.

While definitely a moth, the sphinx flies, maneuvers, and feeds like a hummingbird. Its favorite food is the nectar inside tobacco blooms. There would be no way the sphinx could get at the nectar in these narrow, deep-throated blooms if it did not have the ability to approach the flowers like a hummingbird. The sphinx hovers over the flower while inserting its long tongue into the flower. Its tongue, which is actually longer than the rest of its body, has two grooved halves, which, when fit together, create what amounts to a long straw to draw out nectar. If the two halves don't fit perfectly, the sphinx will starve to death.

Obviously the tongue of the first sphinx moth had to be fully-formed! As it hovers, the sphinx actually rivals the hummingbirds' 50 wing beats per second with its own wing beat of 25 to 45 times per second!

The wonderfully varied patterns in creation do not speak of relationships forged by millions of years of evolution. Rather, they speak of creative relationships, carefully designed by one all-wise Creator!

Prayer: Dear Lord Jesus Christ through Whom all things were made, I pray that I may always be led to give You praise and thanksgiving for all Your wonderful works before men. Amen.

The Shaggy Wasp

Genesis 3:21
"Also for Adam and his wife the LORD God made tunics of skin, and clothed them."

The fiery red and black shaggy coat of the so-called velvet "ant" does more than warn the unsuspecting that it packs a powerful sting. Its shaggy coat protects it from the heat of its desert-like home.

The insect which is popularly called the velvet "ant" is actually a wingless wasp which is often found in hot, sandy, desert-like dune areas. Even though the wasp is unable to fly from the hot sand, it seems to carry on its life with little notice of temperatures which can reach 130 degrees. The wasp's insulating coat of fur helps to keep the heat away from its body.

Many creatures — even plants which live in this hot environment — have protective fur coats often tinged with white to help reflect heat. Digger wasps, which live in the same environment, even have tiny silvery hairs between their eyes in order to protect their tiny brains from the heat. It is from this environment that one popular landscaping plant, the dusty miller, comes with its silvery protective coat covering its normal, green leaves.

The clothes and coats we wear for protection, like those in the rest of the living creation, are but weak images of our ultimate need for protection. When our first parents sinned, God was first to kill animals in what amounted to the first sacrifice for sin to provide coverings for Adam and Eve. Yet these, too, were only pointers toward the complete sacrifice for our sins by Jesus Christ. Through Him we have a perfect and sinless white cloak covering our sins.

Prayer: Dear Father in heaven, I thank You that as You provided coverings for Adam and Eve after they sinned, You have also provided the perfect white cloak of Christ for my sins. In Jesus' Name. Amen.

Peru's Marvelous Hummingbird

Isaiah 28:5
"In that day the LORD of hosts will be for a crown of glory and a diadem of beauty to the remnant of His people."

In 1835, when scientists first saw Peru's most unusual hummingbird, they were so overcome with its beauty that they gave it the name "Marvelous." When one modern scientist saw the Marvelous hummingbird for the first time, he said that he could hardly believe his eyes.

The Marvelous hummingbird treats the eye to iridescent green, yellow, orange, and purple feathers. But its most unusual feature is its tail feathers. While most birds have eight to twelve tail feathers, the Marvelous hummingbird is unique in having only four. Two of these are long, pointed, thorn-like feathers which don't seem to help much in flying or landing. The other two feathers are truly marvelous. They are six inches long, three times the length of the bird's two-inch body. On the end of these two long narrow feathers are large feather fans which nearly equal the surface area of its wings.

Astonishingly, the Marvelous hummingbird has complete control of these feathers. At rest, the bird perches with these two feathers hanging down an inch or so from its body, and then crossing them until they are horizontal. In flight and landing they provide remarkable maneuverability. During mating, the hummingbird moves them as semaphores. Interestingly enough, evolutionists admit that they are stumped as to why these unusual feathers should have evolved.

One look at our creation clearly shows that our Creator appreciates beauty. But even the beautiful Marvelous hummingbird is but a poor and cloudy hint of the beauty of our Creator Himself!

Prayer: Dear Father, help me treat the beauty You have created as You would have me to do. Let me be filled with thanksgiving to You for it, and let it remind me that You are the source of all that is truly beautiful. In Jesus' Name. Amen.

The Giraffe's Wondernet

Job 9:10
"He does great things past finding out, yes, wonders without number."

Did you ever stand up a little too quickly and get dizzy? It can happen because by standing, you may have temporarily lowered the blood pressure in your brain. Can you imagine what can happen when the giraffe swings its head from the ground to a tree-top?

The giraffe's heart and the rest of its cardiovascular system is very different from ours — and from most other creatures. If it weren't different, there wouldn't be any giraffes! In order to get blood all the way from its heart up that long neck to his brain, the giraffe's heart must produce twice as much blood pressure as would be expected in an animal its weight. The giraffe's brain is very sensitive to high blood pressure. So there is a special structure which has been called the "wonder net" where the blood supply enters its brain.

The wonder net keeps the blood pressure in the giraffe's brain just where it should be. Even if the giraffe quickly drops its head, say from nibbling a tree-top to take a drink of water, the blood pressure in its head remains safe. The wonder net quickly controls such rapid changes. And to prevent used blood from draining back into its brain when he lowers its head, the giraffe has a special set of one-way check valves in its neck.

When we talk about our Creator we need not be afraid that some people will think we are talking about worthless things. Truly the wonders which the Creator has made are great and worthy to be told to people far and wide!

Prayer: Dear Lord Jesus Christ, give us the bold conviction we need to boldly tell others what You have done — from our creation to our salvation. In Jesus' Name. Amen.

The Tool-Using Bird

Genesis 3:23
"... therefore the LORD God sent him out of the garden of Eden to till the ground from which he was taken."

At one time scientists who believe in evolution said that one of man's unique characteristics was that he was the only animal to use tools. Of course, those who believe in evolution like to speak of man as an animal so they can ignore his unique spiritual reality and responsibilities. But soon, scientists began discovering a growing number of tool-using animals.

One of the latest creatures to be added to the growing list of animals who use tools is the Egyptian vulture. About the size of a raven, the Egyptian vulture loves to eat the eggs of other birds. To get at the treat inside the shell, the Egyptian vulture will search out a stone to throw onto the egg. While these stones are usually just large enough to do the job, the vultures have been recorded throwing stones which weigh over a pound. They hit their target about 50 percent of the time. While studying the Egyptian vulture, researchers found that the vulture will go after anything that is egg shaped, even if it is much too large to be an egg. But they will totally ignore other objects designed to fool them.

The fact that there are animals which use tools shows that the evolutionary definition of man totally fails to describe what we really are. Tool-using is not some evolutionary development. According to the Bible, when Adam and Eve were sent out of the Garden of Eden, they were given the task of tilling the ground. This shows that man has always been a tool-user and maker.

Prayer: Dear Lord Jesus Christ, so many in our confused and lost age are searching and confused, especially since they have been told that they are nothing more than animals. Raise Your people up so that we can make a meaningful witness to those who are searching. Amen.

Nobel Prize Winner Says Man is Divine Creation!

Rom 1:20
"For since the creation of the world His invisible attributes are clearly seen, being understood by the things that are made, even His eternal power and Godhead, so that they are without excuse."

When he was 17, medical student John Eccles began asking questions about what man is, what his thought is and what the meaning of life is. While teenagers typically ask such questions, most teenagers don't go on to become one of the world's greatest experts in how the human brain works. Nor do they win Nobel prizes for their work. Today Sir John Eccles, who has also been knighted, speaks out about the results of his studies.

Eccles makes clear that the purely materialistic and evolutionary definition of man simply isn't supported by science. He points out that his research has helped us describe in great detail what happens in the brain, nerves, and muscles when you decide to move your finger. But, he says, there is no material event which describes how this complex sequence of actions gets started. Man, he concludes, is more than tissue. Man has a non-material mind. He thinks scientists who say that man can be fully explained by materialistic principles are spreading a modern form of superstition. Their beliefs, he says, are worn out. In his words, "They lead us no where. Materialism gives you a hopeless, empty life, one without values."

While Eccles believes that evolution might explain how our bodies got here, he points out that science shows that this explanation isn't enough. He says, "Each of us is a unique, conscious being, a divine creation. . . . It is the only view consistent with all the evidence." At least to this last one point we say a hearty "Amen!"

Prayer: Dear Father in heaven, I pray that You would provide a witnessing Christian at the side of every person who has been led to an empty life by the materialism of our age so that they might be guided to a full life in Your Son, Jesus Christ. In His Name. Amen.

Swimming in Sand

Romans 11:33-36
"Oh, the depth of the riches both of the wisdom and knowledge of God! How unsearchable are His judgments and His ways past finding out! For who has known the mind of the LORD? Or who has become His counselor? Or who has first given to Him and it shall be repaid to him? For of Him and through Him and to Him are all things, to whom be glory forever. Amen."

Their crystalline world might seem like another planet as they swim past monolithic crystals larger than themselves and down channels lined with nearly clear, often form-fitted boulders. But this exotic world is not on another planet. It may be below your feet.

While the wet sand of the beach may sometimes appear to be completely lifeless, it is literally saturated with life. Under a microscope, you would see countless creatures, some of them so small that they can actually swim in the small amounts of water between grains of sand. There might be tiny sand fleas which were washed ashore, appearing as giants next to the single-celled dinoflagellates which swim freely between the grains. The microscope can also show tiny marine worms and crustaceans of unimagined shapes navigating channels which may be smaller than a human hair. And even if you were to take dry sand from high above the water line and add water, you would soon find it teeming with the same kind of life.

Man's sin has caused untold destruction of the perfect form of God's creation. However, that original creation was so vibrant and wonderful that even the present shadow of what God originally made is saturated with life and beauty. While even our best science can't hope to create anything like this, who among us creatures could even imagine creating such wonders!

Prayer: Dear Lord Jesus Christ, the wonders which were made through You are beyond my ability to imagine. Help me to always remember that I look toward a new heavens and Earth, even as I have been made a new creation through Your saving work. Amen.

Is the Lincoln Memorial Thousands of Years Old?

Matthew 3:9
"'and do not think to say to yourselves, "We have Abraham as our father." For I say to you that God is able to raise up children to Abraham from these stones.'"

Did you know that according to one standard method of dating, the Lincoln Memorial in Washington D.C. was built before the time of Christ?

Anyone who has ever visited a cave has heard the claim that the stalactites growing in the cave take, on the average, a full century to grow only one inch. Cave visitors stand in awe as they view stalagmites and stalactites which are ten, twenty, and even dozens of feet tall. How incredibly obvious it is that the Earth is millions of years old! But we need to ask, does it really take a century to grow one inch of stalagmite? What does experience show? One good place to document human experience with stalagmites is in the basement of the Lincoln Memorial in Washington D.C. When the memorial was built, engineers sank steel cylinders into the bedrock 50 feet underground in order to anchor the monument. The base of the memorial is set high above ground, leaving a cavernous basement beneath the floor. Even though the memorial was built only 55 years ago, water seeping through the marble floor has formed stalagmites up to five feet long in the basement!

In a very real sense, the stalagmites in the basement of the Lincoln Memorial are rocks which are carrying the testimony of God against the naturalistic, long-age stories of evolution. Isn't it only right that we who know His love in Christ should be as bold to witness the true message of His Holy Word?

Prayer: Dear Father, I thank You that You have given me Your Word — the Bible. Help me to truly live as Your child by bearing witness to Your truth in this doubting world. In Jesus' Name. Amen.

Origins: Your Answer Matters!

Hebrews 12:1
"Therefore we also, since we are surrounded by so great a cloud of witnesses, let us lay aside every weight, and the sin which so easily ensnares us, and let us run with endurance the race that is set before us...."

What makes something right or wrong? Are things that are sometimes wrong *always* wrong?

Most writers and commentators, even most preachers today have problems with the question of right and wrong. That seems amazing since only a generation or two ago few people had any problems at all with the issue of right and wrong. About the only act that seems to be universally accepted as wrong today is to claim that there is an absolute standard of morality!

How did things change so fast? That's not hard to answer. When people accepted that they were made by a Creator, they accepted one important consequence of that fact — the Creator owns us lock, stock, and barrel, and therefore has every right to hold us accountable for our actions. Many people understood that God was determined to steer us from the wrong and urge us toward the right because the wrong hurts us while the right provides us with a sense of fulfillment. So, life cannot be full when the Creator is not included.

When evolutionists convinced much of the population that this personal Creator was a myth, people began to reason that without a Creator, right and wrong were up for grabs — morality was not absolute. And that is just what the first architects of evolution said would happen if evolution was adopted!

Prayer: Dear Lord Jesus Christ, it is because You have carried the punishment for my sin on the cross, and because You were raised on the third day that I have a new, meaningful life to live. Help me to live it and so have joy. Amen.

Animals Which Make and Use Tools

Proverbs 3:7-8
"Do not be wise in your own eyes; fear the LORD and depart from evil. It will be health to your flesh, and strength to your bones."

At one time scientists who believed in evolution — always eager to define man as just an animal — defined man as the only tool using animal. But then they discovered that a number of animals use tools. So they changed their definition of man to say that man was the only animal which *makes* and uses tools. Before long, however, scientists began discovering that many of these tool-using animals actually make their own tools.

For example, chimpanzees will make tools out of sticks and grass to help them fish termites out of logs and trees. They also use leaves as humans use facial tissue. Chimps chew leaves to make a sponge which comes in handy for soaking water out of hard-to-drink-from places. Gorillas have been seen using crooked sticks to pull ripe fruit into reach. But notice that more non-primates than primates make and use tools. Elephants use sticks for back scratchers. The California sea otter uses a stone to break open clam shells. The Galapagos woodpecker finch uses a stick or cactus spine to fish for grubs in trees. The satin bower bird makes a paint brush out of bark to paint the inside of its bower.

Our growing knowledge of the gift of intelligence which God has given to the animal world clearly shows that tool-making and using have nothing to do with any supposed evolutionary development. But it does lead us to stand in wonder and thanksgiving at God's goodness toward the creation!

Prayer: Dear Heavenly Father, help me to understand that intelligence, ability and even understanding itself are all Your gifts to us, and not our own accomplishments. Help me to avoid pride and to use Your gifts as You would have me use them. In Jesus' Name. Amen.

Evolutionist Says: Darwinism is in Trouble!

Psalm 104:30-31
"You send forth Your Spirit, they are created; and You renew the face of the earth. May the glory of the LORD endure forever; may the LORD rejoice in His works."

In his book **The Neck of the Giraffe** Dr. Francis Hitching writes that Darwinism is in a lot of trouble. He laments, and I'm using his own words, that evolution "has not, contrary to general belief, and despite very great efforts, been proved."

Hitching admits, to his great sorrow, that fossils do not show any history of evolutionary development. To use his words again, creatures "come into the fossil record seemingly from nowhere — mysteriously, suddenly, fully formed and in a most un-Darwinian way." What's worse, he says, is that the gaps in the fossil record will never be filled with evolutionary ancestors. These systematic gaps show a pattern which cannot be explained away by saying that someday scientists will find the missing creatures. He admits that in the history of life on Earth, plants and animals must be treated as though they came into existence fully-formed in the forms we know today. In other words, biology must assume the creationist explanation. Hitching notes other problems as well. We have no idea how the genetic code could have formed without a Creator, and mutations can't explain the supposed changes of evolution.

I don't want you to make the mistake of thinking that Hitching is a creationist. He argues that there can be no debate that evolution actually took place. After all, he says, we are here and that is proof enough. But his statements are an admission that there is no explanation of how evolution could have happened, and that evolution has no support from scientific fact!

Prayer: Lord God, Your Hand is not hidden, even from those who do not want to see it because Your glory is so great. For this I thank You. And I pray that Your Hand may be even more evident in my life. In Jesus' Name. Amen.

Luring with Light

2 Corinthians 11:14-15
"And no wonder! For Satan himself transforms himself into an angel of light. Therefore it is no great thing if his ministers also transform themselves into ministers of righteousness, whose end will be according to their works."

The spider's web is not simply a net which catches insects that accidentally fly into it. Scientists have recently learned that spiders actually make their webs into both traps and lures using ultraviolet light.

Unlike humans, most insects can see ultraviolet light. They use this ability to locate the sun when it is behind clouds as well as to locate certain flowers which offer rich ultraviolet hues to those who can see them. Spider silk, researchers have learned, is also able to reflect ultraviolet light. This is especially true of silks used for web-making, as opposed to silks created by the spider for lining burrows or protecting eggs. The garden spider even weaves decorations into its web that increase ultraviolet reflection. Studies with insects show that they are lured into the web by the ultraviolet light. In fact, webs that have extra decorations which increase ultraviolet light reflection increase the garden spider's catch by 58 percent. Scientists also think that extra ultraviolet light reflective decorations in the web serve as a warning to birds, who can also see ultraviolet light.

Just as the spider's web is as much a lure as it is a trap, the devil too, does not wait patiently for us to fall into his traps. He lures us to them with things which will catch our interest — even things which appeal to us. This knowledge of how he works can help Christians avoid his traps. But we thank God that He sent His Son to purchase our salvation from sin, death and the devil.

Prayer: Dear Lord Jesus Christ, the devil is so much smarter than I am. I ask that even as You have redeemed me with Your holy, precious blood, You would help me to see the devil's lures and traps and be warned away from them. Amen.

Intelligent Patience

2 Peter 3:9
"The Lord is not slack concerning His promise, as some count slackness, but is longsuffering toward us, not willing that any should perish but that all should come to repentance."

A recently released study on patience and reward echoes what the Bible has to say about patience, reward, and temptation. This is a case of the Bible showing that this study is on the right track.

In the study, researchers offered four-year-old children a promise of a couple of small cookies right away, or if they could wait 15 minutes, five pretzels. The children were allowed to see each treat, including the fact that the pretzels amounted to a much larger treat, when the offer was made. The children were also told that they could press a buzzer at any time before the 15 minutes was up and immediately receive the smaller treat. Researchers left the treats in full sight of the children in some cases, and hid them from view in other cases. After the study, researchers examined how well the various children did in school for the next ten years. They concluded that those children who waited patiently for the larger treat as four-year-olds also did better in school than the children who could not wait. They also found that talking about the treats during the fifteen-minute waiting period caused more of the children to give in to temptation and abandon patience. Researchers suggested that teaching children self-control will help them in school and in their social relationships later in life.

The Bible tells us not to dwell on things we are tempted toward, but to flee even the opportunity for temptation. Scripture also tells us that the Lord is not slow to return to Earth for the same reasons that men put things off. It is His desire that through His slowness as many as possible might be saved.

Prayer: Dear heavenly Father, I am often not a very patient person. I ask Your help to learn to be patient with those around me, even as You are patient with me as well as those who are yet unsaved so that no one may be unnecessarily lost. In Jesus' Name. Amen.

The Lord of History

Acts 17:26-27
"'And He has made from one blood every nation of men to dwell on all the face of the earth, and has determined their preappointed times and the boundaries of their habitation, so that they should seek the Lord, in the hope that they might grope for Him and find Him, though He is not far from each one of us.'"

All of history flows like a river, unstoppable by any man, from the creation to the judgement of all the earth. And every drop of water in that river is not labelled H_2O, but Jesus Christ.

It has been said that God created man in His image, and ever since, man has been trying to return the favor. Man doesn't like to have to look up to a Creator. In his rebellion, man may try to transform God's creating Hand into the dice of chance. But dice cannot create what mind and fingers make. Man can try to say that he can't find the manger with a microscope. Or, he can say that Christ could not rise from the dead because even our smartest scientists can't rise from the dead. Man can claim that the educated have risen above a so-called Christian "religious myth" — but he cannot deny that when he does so, he creates a new religious myth.

All of history finds its meaning in Jesus Christ and all the fixtures of history are covered with the fingerprints of God. So long as man is allowed to be free to see, that fact will never be hidden by the proud claims of any of man's science.

Men may deny that God is in control of all history or that Jesus Christ is the theme of all history. They can try to pollute the river of history or even hide it from view. But man's history is still numbered in years before and after Christ. He, and no man, is the true Lord of history.

Prayer: Dear Lord Jesus Christ, despite the schemes of men to lure people away from You through the senses, so-called science, or apathy, I thank You that You remain the Lord of history Who cannot be denied. Amen.

The Largest Creature on Earth

Genesis 1:21
"So God created great sea creatures and every living thing that moves, with which the waters abounded, according to their kind, and every winged bird according to its kind. And God saw that it was good."

Many people think that the greatest of the dinosaurs were the largest creatures ever to live on Earth. But while a few of the very largest dinosaurs may have been longer than the blue whale, the blue whale has the largest sheer body size of any creature that ever lived.

The blue whale can reach a length of 100 feet and weigh up to 170 tons. That's about how much a whole town of 2,700 people weighs! The blue whale has seven stomachs. When you consider that a blue whale has to eat a million calories a day — the equivalent of 1,000 banana splits — having seven stomachs makes sense. The tongue of a full-grown blue whale weighs more than an elephant. And its arteries are big enough for a man to swim through. Its heart, which has to pump the whale's entire supply of eight tons of blood, weighs 1,000 pounds. And the blue whale can live for 120 years. No dinosaur could ever boast of such statistics!

The whale is a wondrous and magnificent creature. In addition to its great size, it is intelligent. Besides showing forth God's great creative power and imagination, the whale also shows us that neither size nor other standards of human measurement give us an indication of how God values things. Though we human beings are not the largest, nor the only intelligent creatures on Earth, we are the only creatures who are responsible to our Creator for the way we live. That is why, when sin separated us from our Creator, He sent His only Son to save and restore us to Himself.

Prayer: Dear Heavenly Father, help me not to measure things in a worldly manner. Instead of focusing on the creation, including myself, let my focus be on You and what You have done for me. In Jesus' Name. Amen.

The Truth About Those Carbon Dates

Romans 2:14-15
"... for when Gentiles, who do not have the law, by nature do the things contained in the law, these, although not having the law, are a law to themselves, who show the work of the law written in their hearts, their conscience also bearing witness, and between themselves their thoughts accusing or else excusing them."

Anything that was once alive or that was produced by a living thing can be dated by using what is called the radio-carbon method of dating. Radio-carbon dating relies on the fact that all living things take in carbon, some of which is very slightly radioactive. At Bible-Science, we are often asked just how reliable this method really is.

Professor Robert Whitelaw is the leading creationist expert on carbon dating. The method itself was invented in the 1950's by W.F. Libby, a committed evolutionist. Dr. Whitelaw began studying carbon dating results in the 1960's, and has now reviewed over 30,000 carbon dating results. He points out that Libby knew, from his own research, that carbon-14 dating proved that the Earth was only a few thousand years old. But Libby rejected this result as being contrary to his religion — evolution.

When carbon dating information is adjusted to fit Libby's own data, carbon-14 dating demonstrates that there was a world-wide cataclysm which destroyed all life at just about the time given in the Bible for the great Flood. Carbon-14 also shows that all living things appeared at about the same time.

Because they are only human, scientists, like everyone else, are able to find excuses for being less than honest with the facts. As several evolutionists have stated, there is a good case for creation, but they admit that they do not want to accept the idea that there is a God.

Prayer: Dear Lord Jesus Christ, I ask that You would confuse the efforts of those who are trying to hide the truth about you. Help me always to be honest with the truth. Amen.

Birds Helping Birds

Luke 10:33-34
"'But a certain Samaritan, as he journeyed, came where he was. And when he saw him, and went to him and bandaged his wounds, pouring on oil and wine; and he set him on his own animal, brought him to an inn, and took care of him.'"

Altruism — helping one another — seems to be a real mystery to those who explain the world and everything in it in materialistic terms. Since evolution is supposed to work on selfish principles, creatures, including man, have no reason to develop a helping attitude.

Because they believe that everything can be explained by materialistic principles, evolutionists wonder how the genetic code could have first been programmed to produce creatures which help one another. Researchers have just completed their studies of a bird called the white-fronted African bee eater. Members of this species help each other even at the obvious sacrifice of their own welfare. One bird will face spitting cobras, hunt tirelessly for food, and even put off having its own young to help another member of its species. Scientists note that according to the principles which should favor the evolutionary development of this bird, such behavior is suicidal. Researchers have tried to explain the bee eaters' common habit of putting off starting their own families to help other bee eaters raise their young by saying that such behavior is limited to birds who are related. But they admit that even adopted orphaned bee eaters will help their adopted parents in this way.

While it is often denied, the direction of this research shows that belief in evolution does indeed affect one's thinking about morals. Evolution suggests a very opposite morality to that which our Creator would have us follow.

Prayer: Dear God, I ask that You would help me so that I would not adopt the selfish and cold behavior of so many in today's world. Let my life be an example which draws others to You. In Jesus' Name. Amen.

Cigars with Wings

Proverbs 26:2
"Like a flitting sparrow, like a flying swallow, so a curse without cause shall not alight."

The small bird known as the chimney swift is also known as the flying cigar because of its long, narrow form. These four- to five-inch birds are so adapted to the air that they have only small, weak feet which are unable to allow them to walk or perch like other birds. In fact, their Latin name even means "without feet."

When swifts do come to rest, they do so against a vertical surface like a wall. They are often found resting inside the protection of chimneys, which leads to their English name, chimney swift. Few other birds can match their flying abilities. Swifts are so at home in the air that they eat and drink while flying, gather their nesting material on the wing, and sometimes even mate while in flight! They are very social birds, sometimes acting as though they had one mind between them. A large flock of swifts may sight a chimney and begin circling it in smaller and smaller circles. Finally, the birds closest to the chimney enter it, followed by the rest. One person reported seeing a flock of 10,000 swifts enter one chimney over a period of 37 minutes!

When David was cursed by Shimei, he realized that the curse would be of no effect because he was innocent of murdering members of Saul's family. In Proverbs 26:2, his son Solomon compares unjust curses to a bird which never lands, meaning that when people speak evil of us, or curse us, if what they say is unjust, their words will have no effect. It is God's wisdom to think of unjust things which are said about us as being like the swift, always in flight and never landing on us.

Prayer: Dear Lord, when people speak evil of me or wish the worst for me and their feelings are not justified, help me to remember that You are all the comfort I need. And when I really do offend someone, give me a repentant heart. In Jesus' Name. Amen.

Science Tries to See

Psalm 139:14
"I will praise You, for I am fearfully and wonderfully made; marvelous are Your works, and that my soul knows very well."

Scientists have tried to tackle the job of creating a machine which is able to see as well as the human eye. In the process, they are gaining a new appreciation for the wonderful gift of sight.

They have learned, first of all, that no computer chip can be made today which could begin to do what the retina does. But, if it could be made, it would have to be on the order of half-a-million times bigger than your retina. One computer scientist has estimated that a computer chip which could begin to do what your retina does would have to weigh in at about 100 pounds! The retina, which is something like a small slip of clingy food wrap, weighs less than a gram. It occupies only 0.0003 of an inch of space.

On the other hand, the scientists' theoretical "seeing" chip would fill 10,000 cubic inches. And while your retina operates on only 0.0001 of a watt of power, the "seeing" chip would require 300 watts of power and a cooling system. And even with all of this, it couldn't see very well. It would only be able to resolve a square area of about 2,000 units of vision, called pixels, while your eye can resolve five times that much! Such a chip would have the equivalent of about a million transistors, while your retina has the equivalent of 25 billion transistors!

Truly, modern science offers elegant testimony to the fact that the eye could not have been produced by evolution — *and* that it could have only been created by a very wise Creator!

Prayer: Truly, dear Lord, I am fearfully and wonderfully made! I thank You for all the senses You have given me. Even though they may not all be perfect, I ask that You would perfect them by leading me to use them to Your glory. In Jesus' Name. Amen.

It's Not a Spider!

Genesis 1:25
"And God made the beast of the earth according to its kind, cattle according to its kind, and everything that creeps on the earth according to its kind. And God saw that it was good."

What has eight legs, fangs, ridges on its underside, and certainly is one of the "creeping things" which God created? If you said "a spider," you are wrong. While many people think of the daddy-long-legs as a spider, it really is not.

More than just the ridges on its underside set the daddy-long-legs apart from spiders. Spiders have two body sections while the daddy-long-legs has only one. Under a microscope you would also learn that while spider fangs have two parts, the fangs of a daddy-long-legs have three sections.

God has also given the daddy-long-legs some very special abilities to protect itself. Since, when it is attacked, it is most likely to be grabbed by one of its long legs, its legs come off quite easily. And a young daddy-long-legs will be able to regrow the missing leg, while getting around very easily on the other seven. Its eyes are placed on the top of its body so that it can see better. It has especially good sight to spot attacks from above — from which most attacks are likely to come. The daddy-long-legs can also spray a nasty smelling gas to protect itself.

God wasn't satisfied in making only one variety of daddy-long-legs. Even here He did not spare His creativity. There are three species of daddy-long-legs. The daddy-long-legs was also given the important job of cleaning the forest floor of dead insects and decaying vegetation.

Prayer: Dear Heavenly Father, I give thanks to You for Your great wisdom in providing for every detail of the creation and its needs, and for doing so in such wondrous fashion. In Jesus' Name. Amen.

Plant Self-Defense

Amos 8:11
"'Behold, the days are coming,' says the Lord GOD, 'That I will send a famine on the land, not a famine of bread, nor a thirst for water, but of hearing the words of the LORD.'"

The death of some kudu, a species of African antelope, has led scientists to discover the amazing way in which the acacia tree protects itself.

The thorny acacia tree is not the usual food of the antelope. But because of overcrowding on game farms, the antelope began eating the leaves of the tree. Eventually the kudu starved to death, even though there were plenty of acacia leaves to eat. Scientists began unraveling the mystery when they learned that when antelope feed on the acacia, the tree begins to produce a chemical called tannin k. The tannin combines with other chemicals in the leaves to make them taste bad. In addition, when the antelope continue to eat the leaves, the tannin accumulates in their digestive systems, impairing their ability to digest food, and eventually they starve to death. Scientists found that the tannin is not normally found in high concentrations within the leaves. But within 15 minutes of leaf damage, the tannin levels in the leaves nearly doubled. They also found that the trees warn each other about browsing animals in the area. When a leaf is chewed, it releases a scent which causes other acacia trees in the area to begin producing tannin!

Many people today are in a situation similar to these African antelope. Though most places in the world have no shortage of Bibles, Bible reading is decreasing. It is primarily where Bibles are hard to get that people appreciate the power of God's Word. God has provided us with plenty of food for our souls. Let's not replace the Word's true nourishment for the soul with the empty food of the world that will starve us!

Prayer: Dear Lord Jesus Christ, You are the Word made flesh. Please help me to remember this truth when I consider my use of Your Word, the Bible in my life, so that I may remember that it is in the Bible that You instruct me. Amen.

Human Feet of Clay

Matthew 5:37
"'But let your "Yes" be "Yes," and your "No," "No." For whatever is more than these is from the evil one.'"

It is unfortunate that many people have the idea that scientists are something more than human. Sometimes it seems as though, for some people, something isn't true unless a scientist verifies it. Even many scientists come to have this picture of themselves.

But many scientists are honest about the fact they are just as human, just as fallible, and just as prone to bias or even dishonesty as anyone else. The 19th century scientist, Charles Babbage — a creationist by the way — did an analytical study of the problem of fraud in science. He found that there are three basic types. Scientists can and do record observations which never took place. A second type of fraud is to ignore those observations which don't fit the average. A third type is called "cooking" — where only the data which fits the hypothesis is used and the rest is ignored. Scientists may resort to any of these forms of fraud to appear more productive, or simply to make a name for themselves.

Honest scientists also admit that scientific data which fits their own personal view of the world is easier to see than data which goes against their beliefs.

This means that scientists are just as human and prone to the same motivations in their work as the rest of us. Not all scientists are frauds, just as not all car mechanics are dishonest. But science is not some open door to absolute truth. Nor can it judge the truth of God's Holy Word.

Prayer: Dear Father, I face many temptations to be "just a little less than honest." I ask Your help that I may always think and deal with others honestly, recognizing that You are the only Source of all truth. In Jesus' Name. Amen.

Were You a Fish

Romans 1:25

"... who exchanged the truth of God for the lie, and worshiped and served the creature rather than the Creator, who is blessed forever. Amen."

A century ago, a scientist and strong supporter of Darwin named Ernst Haeckel, got into trouble for faking evidence. He supposedly showed that developing embryos go through the earlier evolutionary stages of their species. The scientific community investigated his so-called discovery, and found that he had faked his drawings. Unfortunately, textbooks continue to present this faked evidence in support of evolution.

As a result, many people still believe this false scientific theory. One of the most widely-used "evidences" used to support evolution today is the claim that the developing human fetus has "gill slits" at one point in its development. In fact, new textbooks approved in 1992 in California still presented this claim as a "fact of science."

In actuality, a developing fetus has folds of skin around its neck that contain the developing organs found in the neck. These folds have nothing in common with gills. While no scientist accepts this theory any more, it can still be found even in textbooks used by medical students! Why? It would seem that few scientists have the nerve to challenge this idea which they know is wrong because it might make them look like they are against evolution! And so this anti-scientific idea continues to be used as a "scientific" justification for abortion.

Yes, it's true. The wrong view on origins can result is some disastrous "science" as well as human suffering!

Prayer: Dear Father in heaven, it is especially easy even for Your people to fall into an inadvartant worship of the creation in place of the Creator in our materialistic age. Help me to root this sin out of my life. In Jesus' Name. Amen.

Flagellum Fascination

Luke 8:8
"'But others fell on good ground, sprang up, and yielded a crop a hundredfold.' When He had said these things He cried, 'He who has ears to hear, let him hear!'"

It has only been within the last 30 years that scientists have had the powerful electron microscope with which to probe ever more deeply into the secrets of God's creation. One of the most surprising discoveries they have made with this technology shows an incredible similarity between those little whips called flagellum — which some microbes use for swimming — and your ear!

The electron microscope has revealed that there is but one master pattern for all flagellum. These tiny threads are always made up of a pair of tiny fibers running inside a tube. The pair of fibers is surrounded, inside the tube, by nine additional fibers. It is thought that the contraction of these protein fibers causes the flagellum to move. But though these tiny structures are identical, different creatures use them in different ways.

Inside your ear, and the ears of all vertebrates, there are tiny hair-like structures which come together in a point, something like an Indian tepee. Each of these structures has on it one projection exactly the same as the microbes' flagellum. But in this case, its purpose is not to move, but to be moved. These structures are hooked up to nerves and their purpose is to detect sound waves. They are so sensitive they can provide your brain with many different kinds of information, all of which you finally hear as sound!

Have you used this generous gift of hearing to hear the Word of God today?

Prayer: Dear Lord Jesus Christ, I give thanks to You for all my senses and abilities and ask that You would, through the teaching of Your Word, help me to make use of them in ways which glorify You. Amen.

Leading Evolutionist Becomes a Creationist!

1 Kings 18:21
"And Elijah came to all the people, and said, 'How long will you falter between two opinions? If the LORD is God, follow Him; but if Baal, then follow him.' But the people answered him not a word."

In 1969 Dr. Dean Kenyon, now professor of biology at San Francisco State University, published a book which left him recognized as one of the leading evolutionary scientists. His book, **Biological Predestination**, tried to explain how simple living things could, over millions of years, become more complex, and finally evolve into human beings.

When Dr. A.E. Wilder-Smith, a creationist, wrote his book, **The Creation of Life**, he spent a good deal of time showing what was scientifically wrong with Dr. Kenyon's evolutionary book. Dr. Wilder-Smith is a biochemist with a world-wide reputation. One day one of Dr. Kenyon's students gave him a copy of Dr. Wilder-Smith's book.

After reading Dr. Wilder-Smith's scientific arguments, Dr. Kenyon said, "I found myself hard-pressed to come up with a counter-rebuttal." This was a turning point in Dr. Kenyon's personal search, as he learned that, "It is possible to have a rational alternative explanation of the past." Today Dr. Kenyon speaks for creationism, even offering his students scientific evidence which supports creation — along with teaching them evolution.

We give thanks to God for the honesty in which Dr. Kenyon has dealt with origins. Too many people stand before the truth in silence, hoping they won't have to give up their own ideas. May God make all of us as honest with the truth as Dr. Kenyon!

Prayer: To be honest, dear Lord, there are indeed times when Your truth shakes some of my pet ideas too. I pray that You would grant me Your Holy Spirit so that I may always receive Your truth with thanksgiving, even when I must change my thinking. In Jesus' Name. Amen.

The Octopus with a Shell

Genesis 1:21
"So God created great sea creatures and every living thing that moves, with which the waters abounded, according to their kind, and every winged bird according to its kind. And God saw that it was good."

While the nautilus looks like a sea snail, it actually is much more like the squid and the octopus, and very little like sea snails, except for its shell. Despite the fact that the nautilus shell is found in the very oldest fossil layers which show evidence of life, it has a very concentrated nervous system and well-developed sense organs. The nautilus is found in the southwestern Pacific and eastern Indian oceans, and looks just like the first nautilus recorded in the fossil record. There is absolutely no evidence that the nautilus, one of the oldest life forms in the fossil record, has evolved since life first appeared on Earth.

Like the squid and the octopus, the nautilus has mouth, eyes, and tentacles, and can swim backward quite rapidly by shooting jets of water. It also builds itself a shell home in which to live. As it grows, needing a larger compartment within its shell, it moves forward in its ever-widening shell and closes off the cramped quarters behind. But unlike the average shell fish, the nautilus leaves a fleshy tube of its body in the old "room." This tube extends all the way back to the first room it lived in. This fleshy cord is used to make the shell buoyant by flooding chambers or forcing water out of them. The nautilus is a swimmer, not a crawler!

God's creativity seems to be unlimited as we see more of the amazing ways He has chosen to design His creatures. The nautilus is one more example of the fact that there are no limits to God's thoughts or abilities!

Prayer: Dear heavenly Father, help me to come to You in prayer in all situations. When I am in need, help me to remember to cast all things on You, for Your power, thinking, and creativity are truly unlimited. In Jesus' Name. Amen.

Scientists Distrust Dating Methods

Exodus 20:11
"For in six days the LORD made the heavens and the earth, the sea, and all that is in them, and rested the seventh day. Therefore the LORD blessed the Sabbath day and hallowed it."

To hear some people tell it, scientists have nearly absolute confidence in the dating methods they use. When their dating methods say that something lived 30,000 years ago, give or take 10 percent, they sound pretty certain. On the basis of these methods many scientists announce that the Bible's record of history simply is not accurate. But this is not really how it is when scientists are working in their labs.

In 1966 scientists found, among some fragments of mammoth and bison bone in the Yukon Territory of Canada, a sharpened hide scraper, made of caribou bone. This was clear evidence of man's activity, so they wanted to find out how old the bone scraper was. Radiocarbon dating told them it was 25,000 to 32,000 years old, placing it 20,000 years before man's supposed first arrival in North America. Twenty years later, using a nuclear accelerator, another scientist tested the bones using a different method, and dated the bones not at 25,000 or 32, 000 years old, but at only 2,000 years old!

This is only one example of why scientists often don't trust radiocarbon dates, and why they often refuse to accept them unless they are also dated using the second method.

When God's Word tells us about days and dates, we can trust His Word to be accurate. The Bible is not only a higher authority than any of man's schemes, but has always proven to be unerringly accurate, even when it speaks about history and science.

Prayer: Dear Lord Jesus Christ, You are the Word through Whom we were created made flesh for our salvation. You are the beating heart of Scriptural truth. Help us to see that, for Your sake, all of Scripture is trustworthy. Amen.

Skeptics Admit: Science Can't Deny Soul

John 8:56
"'Your father Abraham rejoiced to see My day, and he saw it and was glad.'"

When God made human beings, did He make them with both a material body as well as a non-material part — a spirit or soul? The Bible, of course, speaks of man's spirit and the fact that the faithful will be with God and be aware they are with God after death. But those who are no friend to Christianity have claimed that the idea of the soul is nothing more than superstition.

Earlier you read about Sir John Eccles. He is the scientist who knows more than any other scientist in the world about how the brain works. As a result of his study, he has come to the conclusion that man has a non-material part to him which controls the brain. The brain is the contact point between the material world and the non-material world. Donald MacKay, who teaches in England; Sir Karl Popper, who is famous for his theory of the scientific method; and neurobiologist Roger Sperry, also believe that man has a spirit.

Most of these and a number of other influential scientists — especially those working in brain research — have concluded that man must have a part of himself that science cannot study, but which influences matter. Other scientists go even further, accepting anti-biblical New Age ideas about man's soul.

So, don't ever let anyone tell you that scientists have shown that man is a completely material creature with no spiritual nature. Some of the most knowledgeable scientists in the world don't agree.

Prayer: Dear Father, in Your wonderful way, You have made it evident that man is more than just a chemical accident. Help me to look forward to being with You forever. In Jesus' Name. Amen.

James Clerk Maxwell: Creation Scientist

1 Corinthians 2:5

". . . that your faith should not be in the wisdom of men but in the power of God."

Albert Einstein said that James Clerk Maxwell made greater contributions to physics than anyone except Isaac Newton. Maxwell developed complex theoretical and mathematical explanations for all the forces in the universe except gravity and nuclear forces. He also made scientific contributions in the fields of thermodynamics and mathematics. In other words, Maxwell was a scientist of gigantic proportions who is still greatly respected today.

By today's standards, Maxwell would be called a "fundamentalist." Maxwell lived at the same time as Charles Darwin, and was very aware of evolutionary theory. He felt strongly that evolution was anti-scientific, and wrote a powerful and important refutation of evolutionary writings. He also offered a careful mathematical refutation of the theory that the solar system had evolved from a cloud of dust and gas.

The great scientist Maxwell believed that Jesus Christ is the Savior which God has provided to deliver man from the results of sin — including eternal death. A writing of his, found after his death, states that the motivation for his work was that God had created all things, just as Genesis says. And since God has created man in His image, scientific study is a fit activity for one's life work.

Prayer: Dear Heavenly Father, I pray today for the work of those in science who are convinced that You are indeed the Creator described in Genesis. Though they are opposed by men, bless their work and move more of our Christian young people to follow in their footsteps. In Jesus' Name. Amen.

Can a Day be a Thousand Years?

2 Peter 3:8
"But, beloved, do not forget this one thing, that with the Lord one day is as a thousand years, and a thousand years as one day."

One of our listeners wrote, asking whether a day in Genesis could not be a great period of time, based on 2 Peter 3:8. In 2 Peter 3:8 we read that with the Lord one day is as a thousand years and a thousand years is as one day.

The key to answering this question is to let Scripture, not science or any other human authority, determine what 2 Peter is saying. If you read the verses which surround 2 Peter 3:8, you will quickly see that this passage is talking about God's patience before final judgment. There is nothing in these verses to suggest that they are intended to explain a special usage of the word "day" in Genesis or any other part of Scripture. Second Peter 3:8 is arguing that God can be patient because He is outside time, and time means nothing to Him. It does not address the subject of how God uses the word "day" when He speaks to us in Scripture.

A study of the Hebrew word for "day" in the Old Testament, apart from Genesis 1, shows us that in absolutely every case where "day" appears with a number modifier or the phrase "evening and morning," the context demands that we understand a 24-hour day. Since that is Scripture's rule in every other case, it must be applied to Genesis 1 as well. 2 Peter 3:8 says nothing about making an exception to this otherwise unbroken rule.

We can truly thank God that He has made His Word clear so that it can be understood.

Prayer: Dear Heavenly Father, I thank You that You are patient with us sinners, especially when we are slow to understand Your Word. Continue to teach me and give me understanding. In Jesus' Name. Amen.

The Amazing Mongoose

Psalm 58:3-5
"The wicked are estranged from the womb; they go astray as soon as they are born, speaking lies. Their poison is like the poison of a serpent; they are like the deaf cobra that stops its ear, which will not heed the voice of charmers, charming ever so skillfully."

The mongoose is about three feet long, and weighs about 10 pounds. Its short legs give it great speed, and yet keep it close to the ground so that it is easier to fulfill his duty as a snake fighter. The mongoose's short legs also offer less of a target for snakes. Mongooses are prized in many parts of the world because they keep the snake population down, which often includes some of the most poisonous snakes in the world. Mongooses are often kept as pets.

One of the most amazing features of the mongoose is its ability to walk away from a snakebite. Even if a cobra bites a baby mongoose, the mongoose is not affected by the venom in the least. This is because the mongoose has venom antibodies in its bloodstream and nervous system. Scientists studying mongoose behavior have found that even a mongoose which has been raised by humans and has never seen a snake instinctively knows how to effectively attack and kill a snake on sight.

The mongoose's size, shape and speed all show it to be excellently designed for killing snakes. Its amazing ability to ignore snake venom, and its internal programming that allows the mongoose to be an effective snake-killer and friend to man certainly shows the power and wisdom of the Creator. And if He can make a creature such as the mongoose to protect humans from earthly serpents, we know that He is also able to provide us with eternal salvation from that old serpent, the devil, through His Son Jesus Christ!

Prayer: Dear heavenly Father, Your wisdom in the Creation is but a dim reflection of Your great wisdom in Your plan of salvation. Help me to remember this, especially when I am under attack by the devil. In Jesus' Name. Amen.

How Many Stars Are There?

Psalm 147:4-5
"He counts the number of the stars; He calls them all by name. Great is our Lord, and mighty in power; His understanding is infinite."

Most of us have heard that our sun is an average-sized star. Yet, this average-sized star churns out more energy every second than we could ever use. Less than one-tenth of one percent of its energy falls on the Earth in any one second. If we could harness only one second's worth of that energy, we would have more energy than man has used in his entire existence or expects to use for thousands of years into the future! All this from only an "average" star!

Some stars are so large that if they sat where our sun is, about 93 million miles away, the Earth would be inside them! It is estimated that there are over 1 billion stars in our galaxy. And it is further estimated that there are over 1 million galaxies in space! This means that as incredible as God's power must be to have created the sun, He actually made over a million times a billion stars! And just think, He did it all in one day and didn't even rest the next day. Scripture also tells us that He calls each star by name.

Just as God created incredible variety on Earth, He has created an incredible variety of stars. Some generate huge amounts of energy, but are completely invisible to our eyes. Others spin at speeds up to dozens of times per second. Some stars actually flash on and off like giant space beacons.

Truly His power, understanding, and wisdom are infinite!

Prayer: Dear Father, I am overcome by the incredible power of Your Word which created everything in space. Help me to remember that it is that same powerful Word which You have had preserved in Scripture and use its power in my life. In Jesus' Name. Amen.

The Master Artist

Genesis 2:9
"And out of the ground the LORD God made every tree grow that is pleasant to the sight and good for food. . . ."

We often speak of God's work in creation as if the only thing which generates our wonder about His work of creation is the wisdom and power of His designs. But we can talk about God as more than just the supreme engineer or planner. He is also the supreme artist.

For example, the universe could get along with a lot less color. It surely wouldn't be as beautiful, but there is really little actual need for all those beautiful sunsets and sunrises. Nor does each one have to be different from any others. And while many flowers have colors which attract their pollinators, few flowers actually need color for this purpose. But flowers in black and white certainly would be dull. There are many beautiful deep-sea fish which don't need their bright and beautiful colors at all. Some of these fish live so far below the surface of the water that only a small amount of blue light reaches those depths. Where these fish live, everything always looks a very dim blue. Yet some of these fish have incredibly bright coloration which they and their friends never see. Here we have to ask the evolutionist what value it could have been for this fish to go through the effort of developing its colors.

In Genesis 2:9 we read that when God created the fruit trees, He also made them so they are "pleasant to the sight." God does indeed appreciate beauty, and as the supreme artist of all time, has created works which human artists can only try to duplicate.

Prayer: Dear Lord, I thank You for all the beauty which You have created in the world. I ask that You would help us to gain a better appreciation for it, and find better ways to preserve it. In Jesus' Name. Amen.

The Swiss Army Bee

Genesis 43:11
"And their father Israel said to them, 'if it must be so, then do this: Take some of the best fruits of the land in your vessels and carry down a present for the man; a little balm and a little honey, spices and myrrh, pistachio nuts and almonds.'"

Just about everybody has heard about the famous Swiss army knife. This tool is much more than a knife. Folded into the handle one also finds a bewildering array of other tools, including a bottle opener, can opener, corkscrew, fingernail clippers, file, screw driver, scissors plus much more.

The legs of the worker honeybee are very much like the Swiss army knife. Each leg has an extra joint between the knee and the joint of the foot. Each of these joints is constructed differently from the others. The bee's fore-leg has a special notch with bristles like a brush for cleaning its antenna. When its antennae are dirty, the bee cannot detect smells. The middle leg has what has been called a crowbar which comes in handy for a variety of jobs in the hive. The back legs of the worker honeybee have pollen baskets which the bee uses to carry pollen to the hive. In addition, the back legs have a combination of spears and pincers for use in defending the hive. The back legs also have cleaning bristles for scraping pollen off the middle legs, while the middle legs have the same bristles for scraping pollen off the front legs.

Just as no one finding a Swiss army knife would think that the knife was just a formation created by nature, no one should think that the worker honeybee is something created by nature. God has made the bee's legs for the purpose of making it an efficient link in the production of honey, a food prized by many living things, including man.

Prayer: Dear Lord Jesus Christ, we see everywhere in the creation, that as the Maker of all, You have generously spared nothing to provide us with good things. We thank You for this care. I especially thank You that You even gave up Your life for my salvation. Amen.

God's Superglue

Psalm 119:31

"I cling to Your testimonies; O LORD, do not put me to shame!"

Scientists say that it is not man, but the lowly barnacle that makes what may be the strongest glue in the world. This glue, called *anthropodin* is so effective that it even works under water. It enables barnacles to stick to the bottom of ships, sea turtles, rocks and piers.

Since an accumulating growth of barnacles can slow a ship down, they have to be scrapped — usually every six months. Throughout man's history, he has tried to keep barnacles from slowing his ships. The Phoenicians, who were among the greatest of the ancient seafarers, tried to keep barnacles from sticking to the hulls of their ships by coating the hulls with tar. They quickly learned that the glue produced by the barnacle could even stick to tar! Modern seafarers have learned that there is no stopping the barnacle's glue — at best, they can only slow down the inevitable. Scientists are now studying the glue with the hope of being able to manufacture it for human projects.

Despite man's great claims for his science, everything man makes can be improved through a better understanding of how God did the same thing in the creation. The same holds true, in a sense, in spiritual things. None of us by ourselves clings as tightly to God's revealed Word as we need to. This is why we end up filled with worry, get depressed, or have other problems. Each of us needs to cling to God's Word with a spiritual glue that is even stronger than the barnacle's.

Prayer: Dear Father, I confess that I do sometimes worry, get depressed, and show other signs of not clinging to Your life-giving Word as I should. For Jesus' sake forgive me, and give me the spiritual strength to cling even more securely to Your Word. Amen.

Answering Young Peoples' Questions

Proverbs 22:6
"Train up a child in the way he should go, and when he is old he will not depart from it."

Could it be true that we have unwittingly been telling our young people that Christianity is irrelevant?

As I present the evidences for creation and explain to young people the problems with evolution, a similar pattern emerges in each presentation. While the material is new to them, it rings true. Soon the questions start and they begin to realize they have been misled by those teaching evolution. Before long, though this is not the effect I was working for, many of them begin to feel angry because they have been misled by their teachers.

When I finish, many of them surround me to ask more questions and make comments. The words of one young man sum it all up. He told me, "I just wish that my older brother had heard this message a couple of years ago. I talked with him a lot and I just know that if he knew that the Bible had intelligent answers, he would not have left the church and hurt my parents so. As a matter of fact, until I heard you I was ready to follow him." Other young people tell similar stories. They grew up learning about creation and Adam and Eve in Sunday school. As they were exposed to evolution, they desperately wanted the Bible to offer an intelligent alternative to evolution. But no one could ever offer it to them.

Our silence seemed to confirm what evolution suggests — Christianity is outdated. Could it be that through our young people the Lord is telling us that we cannot be neutral or ignorant about the question of origins?

Prayer: Dear Lord, Your truth is all in all. Give Your people a stronger desire to address the questions being asked today with the answers provided by Your Word so that more may see that the Bible offers an intelligent alternative to the thinking of the world. In Jesus' Name. Amen

Christian Thoughts on a Trip to Mars

Genesis 1:28
"Then God blessed them, and God said to them, 'Be fruitful and multiply; fill the earth and subdue it; have dominion over the fish of the sea, over the birds of the air, and over every living thing that moves on the earth.'"

In July of 1989 people all over the Earth celebrated the 20th anniversary of man's first landing on the moon. During the celebration President George Bush challenged the American people to the effort of landing on the planet Mars. Since it will take nine months to get to Mars, the first people to land on Mars will begin the work of setting up a permanent manned base there. In part, they will use Martian natural resources to exist.

But what should the Christian think about the effort to land on Mars? Is it a waste of effort which could be better used on Earth? Did God intend man to remain on Earth only? In Genesis 1:28 God gave what many Christians call "the great science commission." He told man to rule and make use of the Earth. This can easily be understood to apply to all the creation which God places in man's hand through the desire which He placed within us to explore. While scientists tell us that we will find minerals and other resources in space which are helpful to us on Earth, God has already blessed our space efforts. Bible-Science's *Creation Moments* broadcast, like gospel broadcasts which can be heard all over the world, is broadcast through technology which is a result of the space program.

Christians have been deeply involved in the space effort, and it was because of Christians that the Bible became the first book to travel to the moon. God has blessed our efforts in space and mankind is better for those efforts. It would seem that as long as we continue to approach space as we have, God will use it for good.

Prayer: Dear Lord, I thank You for Your protection for those brave men and women who have been involved in space exploration. I ask that You would continue to move faithful Christian young people to be involved in science so that You can use our space efforts to glorify Yourself. In Jesus' Name. Amen.

Literally Making Yourself Sick

1 Peter 5:7
". . . casting all your care upon Him, for He cares for you."

We have all heard of someone worrying so much about something that they become obsessed with fear. This is often what is referred to when we talk about someone being "worried sick." But medical science is learning that there is even more danger in worry than becoming controlled by fear.

A number of studies have now confirmed that continued grief, worry, or fear can literally make us sick. Scientists are even learning just how this works. A state of worry, fear, or grief causes chemical changes in the brain, pituitary gland, and adrenal glands. These changes ultimately have the effect of weakening our immune system which protects us from disease. As a result, a wandering cold or flu bug or even a skin infection which might be easily handled by our immune system can get out of control.

Continued fear, worry, or grief can also cripple our body's ability to fight cancer — something scientists tell us our bodies are doing all the time. Other changes in blood chemistry during these periods can also increase the risk of stroke or heart attack.

When we worry we are forgetting that it is God, and not ourselves, Who is in charge of the universe and even what happens in our lives. This makes worry a form of idolatry. It is because God loves us and wants the best for us that He invites us to cast all our cares, worries, and concerns — even our grief — on Him.

Prayer: Dear Heavenly Father, I confess that I often forget how intimately involved in my life You want to be. Give me increased faith to cast all my cares on You, and not take them back to myself after a little while. In Jesus' Name. Amen.

Are Dinosaurs Extinct?

Proverbs 30:18-19

"There are three things which are too wonderful for me, yes, four which I do not understand:.. the way of a serpent on a rock...."

Are dinosaurs extinct? Decide for yourself!

The island of Kommodo lies in one of the most remote regions of the 13,000 islands which stretch from mainland Asia to New Guinea. In 1911, Dutch explorers returned to the west with stories about a huge, man-eating dragon that lived on the island of Kommodo. Scientists dismissed their stories as impossible, since dinosaurs are supposed to have been extinct for millions of years.

Eventually, science had to accept the reality of the Kommodo dragon. The huge reptiles grow to a length of 11 feet and can weigh 450 pounds after lunch. They are fast enough in short bursts to catch and bring down a wild horse, a deer or a man. In fact, the Kommodo dragons have killed and eaten seven men in recent years. They are known to enter the small village on the island and steal goats, but they are cannibalistic as well. Their layers of serrated teeth harbor microbes which can easily cause a fatal infection, so they eventually catch up with just about anything they bite. After seeing footage of these creatures, I have to say that they are by far the most ugly, aggressive, and disgusting creatures I have ever seen. If the extinct dinosaurs were at all like the Kommodo dragon, we can all be thankful for their extinction.

Are dinosaurs extinct? It would be hard to find a difference between the Kommodo dragon and extinct dinosaurs.

Prayer: Dear Lord, as Scripture pictures the devil as "that old dragon," help me to picture him and everything associated with him as even worse than the Kommodo dragon, and help me to flee to You. In Jesus' Name. Amen.

Soviet Life Magazine Tolerant of Christian Concepts

1 Timothy 4:15
"Meditate on these things; give yourself entirely to them, that your progress may be evident to all."

The magazine *Soviet Life* was published in the United States by the Soviets under a joint publishing agreement with the United States. This magazine gave Americans multiple pictures of the Soviet Union, and allowed them to read what the official Soviet press had to say. The August, 1982 issue of *Soviet Life* offered some very interesting comments about the biblical account of origins and the great Flood.

An article entitled "Reflections on the Multiplicity of Worlds" marvelled at all the peculiar coincidences which placed the Earth in about the only place in space where conditions are known to be right for life. *Soviet Life* also admitted that even the best of modern science simply can't answer the question of how life formed from non-living material. But then came the most surprising statement. The author said that the great Flood of Noah had been considered a myth until archeological digs in Mesopotamia found sediments offering clear evidence of a large flood. While the writer's dates, and claims that the great Flood was only a local flood can be dismissed on a number of different grounds, it is interesting how the Bible has had enough influence even in the officially atheist former Soviet Union to receive such familiar mention.

Scripture teaches Christians that if we truly live our faith, the excellence of Christianity will be evident to all, even to those who reject it. And through this evident excellence, Christianity will influence human society and culture.

Prayer: Dear Lord Jesus Christ, You have called Christians to be good leaven in the world. Enable me, in the manner of my day-to-day walk, to improve how others view Christianity. Amen.

The Double-Life of the Hummingbird

Psalm 4:8
"I will both lie down in peace, and sleep; for You alone, O LORD, make me dwell in safety."

You might guess that the hummingbird, darting around from flower to flower with wings beating some 60 times a second, must burn a lot energy to keep going. The fact is, the hummingbird will starve to death if it takes longer than two hours to find food. If a 65-pound boy ate in the same proportions as a hummingbird must eat, the boy would have to eat 100 pounds of chicken every day.

If you were a hummingbird and knew all of this, you would probably be too afraid to take eight hours of sleep every night. But the hummingbird does sleep for eight hours every night. How does he do it?

God has given the hummingbird a most remarkable metabolism. During the day the hummingbird's heart must beat 10 times every second as it keeps its incredibly fast metabolism going. But when it goes to sleep the hummingbird's heart slows down to less than one beat per second — about the same as ours. And to further slow his metabolism, the hummingbird's normal daytime temperature drops from 100 degrees to the same temperature as the night air — 50 or 60 degrees. This drop in temperature would kill most warm-blooded animals. But all of this enables the hummingbird to go without food for a good eight-hour sleep.

The hummingbird provides more than enough evidence that the Creator really does care for His creatures, even when they are asleep.

Prayer: Dear Father, I thank You that You care for me even when I am asleep and cannot protect myself. Comfort me with this truth, especially when I am fearful of the night. In Jesus' Name. Amen.

Is There Life on Other Planets?

Romans 8:22-23
"For we know that the whole creation groans and labors with birth pangs together until now. And not only they, but we also who have the firstfruits of the Spirit, even we ourselves groan within ourselves, eagerly waiting for the adoption, the redemption of our body."

Back in 1836, well-known astronomer Sir John Herschel wrote a series of articles for the *New York Sun* reporting his discovery of life on the moon. He told of how he saw buffalo, goats, cranes, pelicans, beaches, forests and even winged batmen. The *Sun* finally admitted that the whole thing had been a prank, and most of the public was amused.

There is as much evidence for life on the moon or other planets today as there was in 1835. And as we explore the other planets of the solar system, the evidence grows that there is little hope that other life will be found in space. Evolutionists, of course, expect life to be found in space, because they feel that if life evolved on Earth, it certainly should have evolved elsewhere, too. At a recent conference on whether there is life in space one of today's most famous astronomers, Dr. Robert Jastrow, remarked that the question was "essentially a religious controversy."

That the question of life in space is religious is an important insight. Many evolutionists believe that the discovery of life in space would be the final nail in the coffin of Christianity and proof of their religion of evolution. It would not be, of course, since the Bible does not offer any clear statements on the matter. Most Bible-believing scholars, however, believe that the Bible implies that material life exists only on earth. What our exploration of space has taught us is that our Earth is a very special place, carefully designed just for life.

Prayer: Dear Father, I thank You for an Earth so beautiful that even after the destruction of sin, the beauty around us naturally makes us think of You. Help me to add my words to that witness. In Jesus' Name. Amen.

The Plant that's as Active as a Hummingbird

Genesis 1:11
"Then God said, 'Let the earth bring forth grass, the herb that yields seed, and the fruit tree that yields fruit according to its kind, whose seed is in itself, on the earth'; and it was so."

The voodoo lily is native to Southeast Asia, although it is sometimes sold in the west as a curiosity. When it flowers, the voodoo lily sends up a fleshy, purple spike which can be three feet long. Besides being one of the worst-smelling flowers you'll ever experience, the spike has another amazing ability.

The spike of the voodoo lily generates a number of hormones which, when heated, smell like rotting meat. It does this because the flies which pollinate the lily feed on decaying flesh. The smell draws them in to look for food, and in the process pollinate the lily. The flower stalk is able to generate plenty of heat which creates its awful smell. Temperatures inside the flowers can reach 110 degrees F. In order to generate this much heat, the voodoo lily actually generates a metabolic rate similar to that of a hummingbird in flight! This frantic rate continues only until the flower is pollinated.

You might be interested to know that while the voodoo lily is the most dramatic of heat-producing plants, several common household plants which are related to the voodoo lily also produce heat when flowering. These include the dieffenbachia and the philodendron.

Scripture offers a very modest account of God's creative activity. Incredibly imaginative and intricate designs are included in the simple statement, "Let the earth bring forth grass . . . ."

Prayer: Dear Lord Jesus Christ, the Instrument through which the creation was made, I thank You for the wonder-inspiring beauty and diversity around us. But most of all I thank You that You cared so much for the creation that when we fell into sin, You gave up Your life to save us sinful human beings. Amen.

Deep Frozen Squirrel

Job 37:8-10
"The animals enter dens, and remain in their lairs. From the chamber of the south comes the whirlwind, and cold from the scattering winds of the north. By the breath of God ice is given, and the broad waters are frozen."

Researchers recently announced that they have discovered a species of mammal which can actually survive being frozen for weeks.

After studying the hibernation pattern of the Arctic ground squirrel, scientists were amazed to find that this creature, which can weigh up to two pounds and grow longer than a foot, allows its body to drop to 27 degrees (F), (five degrees below the freezing point of water), for up to two weeks at a time during its equally unusual eight- to ten-month hibernation period. After the two week period, the squirrel slowly rouses itself, returning its body temperature to normal. It takes care of a few bathroom duties, and then returns to a state of nearly frozen hibernation for another two weeks. The squirrel usually comes out of hibernation for its short summer in June. It has only two or three months before the ground freezes again and it returns to hibernation, so the squirrel is very busy eating for the rest of the year and mating. You could say that the Arctic ground squirrel sleeps most of its life away.

Scientists say that the Arctic ground squirrel is the only mammal which is able to allow its body temperature to fall below freezing. If they can find out how the squirrel does it, they believe the same method might be used to preserve organs for transplant longer than a few hours. So once again, scientists expect to learn new medical methods by studying how the Creator does the same thing.

Prayer: Dear Father, Your understanding and wisdom in designing the creation are so great that even those who do not want to recognize You still expect to learn from You. As they do so, make it ever more difficult for them to deny You. In Jesus' Name. Amen.

He Breathes with His Feet

Job 21:22
"Can anyone teach God knowledge, since He judges those who are on high?"

What breathes with its feet, has thousands of jaws but no mouth, has up to fifty arms, and an eye on the end of each arm? No, it's not a new creature for the next *Star Wars* movie. It's the starfish.

When God designed the starfish, it almost seems as if He tried to see how different He could make this creature from all others. The starfish can have anywhere from three to fifty arms, depending on the species. And one eye is located at the end of each arm. The rough skin of the starfish is covered with what appear to be spines. But these aren't really spines at all. Rather, they are tiny jaws. They cover the starfish's body to keep parasites from attaching themselves to the starfish. No matter where a parasite tries to attach itself to the starfish it's going to be bitten. Even more amazing is the fact that each of these thousands of jaws works independently of the rest. To get its oxygen, the starfish takes water in through tubes in its feet, each containing a tiny pump and a pipe system linking it to the other feet.

The starfish seems to present us with the lesson that our Creator did not have to make the creation in any certain way. This truth is referred to as the "voluntary creation." Everything was up to Him. If He wanted, you and I might be breathing through our feet — although I'm glad I don't. But the biblical truth of the voluntary creation was one of the crucial ideas that provided the basis for the modern scientific method.

Prayer: Dear Lord Jesus Christ through Whom all things were made, I thank You in wonder and awe at Your creativity, and especially that You made me, and purchased and won me from sin, death and the devil. Amen.

Voyager 2's Encounter with Neptune

Deuteronomy 4:19

"And take heed, lest you lift your eyes to heaven, and when you see the sun, the moon, and the stars, all the host of heaven, you feel driven to worship them and serve them, which the LORD your God has given to all the peoples under the whole heaven as a heritage."

As Voyager 2 made its contact with Neptune nearly 3 billion miles from Earth, the reports from scientists at the Jet Propulsion Laboratory in Pasadena, California, created an excitement which we have not experienced since the landing of the first man on the moon.

But did Voyager learn anything that fits in with what we learn in Scripture? Voyager's first message to us should be obvious. *Our Earth is special!* Neptune is hundreds of degrees colder than any temperature ever recorded on Earth. Worse, storm winds on Neptune were measured at over 800 miles per hour!

Might the message of Voyager someday be read by a creature from another planet? The Bible doesn't give us a direct answer. But it does say that man's sin on Earth placed the entire creation in bondage. This strongly suggests that life only exists here on our special planet. The beautiful rings of Neptune also offer support for the Bible's account of creation. If the solar system was billions of years old, the rings would no longer exist. So the rings are evidence of a young creation.

Man may venture into space, but Earth is our home, and always will be. We will always long for that perfect garden in which we were placed by the Creator Who hoped for, and made us for, an intimate, personal relationship with Him. And that means that Man will always need the Good News of Jesus Christ through Whom we are restored, once again, to our wonderful Creator.

Prayer: Dear Heavenly Father, I thank You that I live in such an exciting age in which so many wonderful things are being learned about what You have made. Give me the words which will enable me to use these exciting discoveries to tell others about You and Your love for us in Jesus Christ. Amen.

God Created Humor

Ecclesiastes 3:1,4
"To everything there is a season, a time for every purpose under heaven: ... a time to weep, and a time to laugh; a time to mourn, and a time to dance...."

Life would certainly not be the same without humor. God certainly appears to think so. Just look at some of the humorous designs He created.

One species of ant plugs the entrance to its nest to keep out invaders. Certain select members of the nest are chosen to plug the entry with their heads — and they have heads which look like cork. The caterpillar of the lobster moth tries to terrify its attacker by assuming a threatening position. If that doesn't work, the caterpillar exposes fake wounds to give its attacker the idea that it has parasites. Some spiders pretend to be ants in order to get close to their prey. But since ants have six legs and spiders have eight, the spider holds its two front legs over his head, pretending to have six legs and a standard set of ant antennae. The Amazon leaf fish puts on a similar show. When it spots prey, it acts like a leaf floating in the water, and floats downstream to get close to its intended victim.

Perhaps my favorite is the near eastern lizard who has an ingenious method to keep from being eaten. When an enemy picks up one of these lizards to swallow it, the lizard grabs a stick and holds it sideways in its mouth so that the attacker won't be able to swallow it.

Like everything else, humor is a creation of God. In fact, it would seem that our appreciation of humor was given to us because He appreciates humor. After all, scientists tell us that laughing is good for us.

Prayer: Dear Lord, I thank You for the gift of humor. I realize that because of sin, like everything else, humor can be perverted into something You did not intend, and which is not good for us. Renew my mind so that I may better appreciate wholesome and godly humor. In Jesus' Name. Amen.

The "Stuff" of Life Found in Space?

Psalm 104:24-25
"O LORD, how manifold are Your works! In wisdom You have made them all. The earth is full of Your possessions; this great and wide sea, in which are innumerable teeming things, living things both small and great."

Voyager 2 has now passed Neptune and is headed out into the huge void of interstellar space. Scientists have speculated about whether the chemicals needed to form life might now be working their magic on Neptune or Triton.

Evolutionary scientists speculate that since some of the outer planets or their moons have ammonia, methane, and water vapor, maybe life is forming below the clouds over which Voyager silently flew. This is based on experiments done some years ago on Earth. Researchers placed ammonia, methane, and water vapor in a special container and sparked electricity through it. This experiment was supposed to reproduce what evolutionary scientists thought the early Earth was like. The experiment produced some amino acids, the simplest of the molecular building blocks of DNA. "Simple" is the important word here. Amino acids compare to DNA like a home run compares to a moon shot.

More recent research has shown that the experiment was not realistic. First of all, the early Earth would not have traps to protect the amino acids from additional sparks which would have destroyed them. Research has also shown that the Earth has always had plenty of oxygen, which would prevent the amino acids from forming in the first place. Worse, the experiment always creates a mix of amino acids which are useless to life. All of this is simply the scientific way of saying that it is not possible for life to start by itself without a Creator. And that's just what Genesis says.

Prayer: Dear Lord, as Your Word says, You are the author of life. I pray the testimony of Your creation would speak this fact more clearly as scientists learn more about life, so that they, too, may be led to eternal life in Your Son Jesus Christ. In His Name. Amen.

Can Bacteria Think?

1 Corinthians 9:24
"Do you not know that those who run in a race all run, but one receives the prize? Run in such a way that you may obtain it."

While it sounds silly to ask whether bacteria can think, the fact is, science has known for over a hundred years that the little fellows can indeed think. Experiments in 1883, conducted by Wilhelm Pfeffer, showed that bacteria will swim toward good food like chicken soup and will swim away from poisons like mop disinfectant.

Pfeffer also learned that bacteria make decisions. He made sure that his bacteria knew the location of the chicken soup, then separated them from it with a mild mixture of disinfectant. He found that in this situation they swim as fast as they can through the disinfectant in order to get to the soup.

This is the same kind of decision-making process you and I go through every day when we tolerate the unpleasant in order to arrive at the pleasant.

This is amazing to most scientists and many others today because we have been trained to think of intelligence in terms of evolution. The "higher" or more evolved a creature is, the smarter we expect it to be. But if we recognize, as the Bible says, that all life is the product of an intelligent Creator, we should not be surprised to find that intelligence has nothing to do with supposed evolutionary relationships. Every creature has been given as much intelligence as it needs by the Creator. He truly cares for every living thing — even bacteria!

Prayer: Dear Father in heaven, when it seems that no one cares about me, help me to remember that You care, and that You made me so that You could care about me. In Jesus' Name. Amen.

Your Body's Wiring

Leviticus 23:40
"'And you shall take for yourselves on the first day the fruit of beautiful trees, branches of palm trees, the boughs of leafy trees, and willows of the brook; and you shall rejoice before the LORD your God for seven days.'"

How would you like to design the telephone wiring system for a 100-story skyscraper? You would have to figure out how to place hundreds of miles of wires so that literally thousands of telephone calls could be going on at the same time.

Actually, your body does something like this, involving millions of "telephone calls" every day. The so-called "telephone system" that I'm talking about is your nervous system. Every second, millions of impulses are travelling from all parts of your body, at 150 miles per hour, to your brain and your brain is sending instructions back through the system. Your nervous system is really made up of three different systems so that it can be more efficient. The central nervous system, which is made up of your brain and spinal cord, is the switching center for all nervous activity. The peripheral nervous system links your brain to the furthest parts of your body. The autonomic nervous system controls those jobs in the body which are done without your thought, like your heart beat. Your brain is at the top of this network. It has about 100 billion nerve cells — 20 times as many cells as there are people in the entire world!

Your nervous system is a beautiful testimony to the Creator's skill and His wonderful gift to you. He has given you your nervous system to interact with the world that He made. It is that activity, in the form of walking and talking, the appreciation of His beautiful creation and the touches of loved ones which keeps your nervous system healthy.

Prayer: Dear Father, help me to keep my nervous system healthy through wholesome stimulation and thoughts so that it can always be a source of my thanksgiving to You in all things. In Jesus Name. Amen.

The Scuba Bug

Genesis 1:31
"Then God saw everything that He had made, and indeed it was very good. So the evening and the morning were the sixth day."

The insect called the water boatsman is a marvel of engineering design. And many of his designs could not have gradually developed over a long period of time.

The water boatsman is about an inch long. Like all insects it has six legs, but these legs are not all alike, nor do they all have the same jobs to perform. The water boatsman swims along on its back. So the Creator gave it extra long and extra strong back legs, with little paddles on the end, so that they work just like the oars of a boat. But if the boatsman's eyes were placed as they are in other insects, the boatsman would not be able to see very well while swimming on its back. So its large eyes are placed for perfect sight while swimming on his back.

Much of the boatsman's swimming and living is done under water. But the boatsman doesn't have gills. To solve this problem, God gave the boatsman the ability to breath through its tail. The boatsman simply sticks his tail out of the water. Tiny hairs keep water out while he draws in fresh air rather like refilling his air tanks for another dive. This pose would leave him in a vulnerable position if it were not for the fact that these hairs also have the ability to sense nearby movement in the water. They are so sensitive that the boatsman can even tell whether the movement is made by something which would threaten him or something that might by good for lunch.

The boatsman is elegant testimony to the Bible's witness that living things were first made in finished form by the Creator.

Prayer: Dear Lord, even though the creation has been scarred and damaged by sin, its beauty and wonder is still marvelous to behold. For this I thank You as I look forward to seeing the New Creation. In Jesus' Name. Amen.

Research Surprises Scientists with Bible's Answer

Ephesians 6:1
"Children, obey your parents in the Lord, for this is right."

These days Christian parents are under a lot of pressure to be more "permissive" and less "authoritative" in raising their children. Now the virtues of permissive parenting have been shown to be in error, and the Bible's direction for parents has been confirmed.

In a study of 124 youngsters conducted over a twelve-year period, researchers studied parenting styles and the effect of each style on the development of young people. Researchers divided the parenting styles they saw into categories ranging from parents who didn't seem to care to those who were extremely demanding and restrictive while offering their children little emotional support. They found that parents who consistently set down clear standards of conduct and clearly-defined limits, while offering close emotional support, produce teenagers who fared better in academic tests, are emotionally more stable, and are much less likely to be involved with alcohol or drugs than any other group. The so-called "democratic" parenting style, they noted, produced far more heavy drug users than parents who set limits. They also noted that restrictive parents who offer little emotional support for their children tend to produce teens, who, while not heavily involved with drugs, are less happy with life and fall below average on academic tests.

Researchers concluded that the limit-setting, rule enforcing parents, who offer strong emotional support to their children are the most successful parents. And this is exactly the style of parenting which Scripture teaches us.

Prayer: Dear Father, I ask that You would provide good support and instruction through fellow believers for Christian parents so that fathers would seek to be toward their children as You are toward us. In Jesus' Name. Amen.

Ice Age or Greenhouse?

Genesis 8:22
"'While the earth remains, seedtime and harvest, and cold and heat, and winter and summer, and day and night shall not cease.'"

I recently ran across an article from a popular science magazine which said that climatologists see a trend toward a change in our climate — toward a new ice age. The article notes that "most climatologists believe" that much of the 20th century has been abnormally warm, and we are, since 1960, entering into what is a more normal cool period.

This most interesting article pointed out that the world's climate is already cooler than at many points in recorded history. At one time Greenland was really green, and England was a major producer of wine. The article notes that a few scientists think the Earth is warming, but most climatologists agree that Earth's climate is cooling. Some even say that we may be in for a new Ice Age.

This is one instance in which a ten-year-old article is more interesting than a current article. These dire predictions appeared in the December, 1977 *Science Digest*. The certainty of evidence for a cooling Earth is most interesting in light of the certainty, only a decade later, that the Earth is heating up. And notice how both predictions are working from nearly the same temperature records. The article unconsciously notes the reason for this fear — the winter of 1976-'77 was especially cold and snowy for much of the northern hemisphere. Now, more than a decade later, a warm summer or two has started talk about a runaway "greenhouse effect." Our best response is to remember the Lord's promise to Noah that the seasons, seed time, and harvest, will continue until the end of the world.

Prayer: Dear Lord Jesus Christ, I ask that You would protect me and all who are Yours from the fears which so terrify a world which neither knows or regards Your promises. Help Your people to be lights in the midst of this darkness. Amen.

How Do Plants Know it's Fall?

1 Corinthians 15:36-38
". . . what you sow is not made alive unless it dies. And what you sow, you do not sow that body that shall be, but mere grain; perhaps wheat or some other grain. But God gives it a body as He pleases, and to each seed its own body."

In the autumn much of the Earth's northern hemisphere prepares for winter. Leaves fall from trees as cold air builds over the sections of the continents that are closest to the pole. But did you ever wonder how plants know when to prepare for fall?

It's not just colder temperatures that tell plants to prepare for winter sleep. Many, if not most, measure the length of the day. Nevertheless, many tree buds must be exposed to a certain amount of cold weather before they even consider the possibility of spring. Apple buds need 1,000 to 1,400 hours of near-freezing temperatures before they think about spring.

However, many flower plants actually measure the length of darkness in a 24-hour day. Of course, the longer the dark period, the closer one is to winter's blasts. Many seeds actually depend on winter to crack their seed coats enough to soak up water for sprouting in the spring.

When winter arrives, millions of people look out on a white, barren, seemingly sterile world. It is hard to believe that just a couple short months before all was luxuriant green growth.

It is no less a mystery to all of us that one day the dead will rise and believers will spend eternity with their Lord and Savior, Jesus Christ. Yet life rising so gloriously out of the apparent dead of winter shows a similar power of our Creator on a smaller scale.

Prayer: Dear Lord Jesus Christ, I look forward to the time when You shall return to Earth and raise me and all believers to life once again. Help me each day to prepare for eternity with You. Amen.

Truly Mated for Life

Genesis 2:24
"'Therefore a man shall leave his father and mother and be joined to his wife, and they shall become one flesh.'"

A mile beneath the ocean waves, where it is cold and dark, a fish takes out her fishing pole and goes fishing. But since it is dark, no one sees her "bait." Realizing this, she turns on her little orange light to guide the way.

When an intended victim sees the light, its small size convinces him that it must be a smaller creature — about the right size for his lunch. He moves in to strike at the light, but it jumps just out of his reach. Lunging a second time, he suddenly finds that he has become the lunch as the long sharp teeth of the anglerfish hold him tight.

The anglerfish is about three-and-a-half feet long, and its mouth takes up much of its size. Just over its top lip the anglerfish has a six-inch long fishing pole with that little orange light on the end. When not fishing, the rod fits in a groove on the top of its head.

The strangest part of the anglerfish's life is its mating practices. While the female is over three-feet long, the male anglerfish is only a half-inch long. He has no fishing pole and no light. Mating starts when the tiny male swims up to the female and sinks his teeth into her side. He never lets go for the rest of his life. In fact, soon his skin and circulatory system actually join the female's, and through this union they become, literally, one flesh. Eggs are eventually laid.

Could it be that our Creator provided this arrangement as sort of a reminder for us that there is much more to marriage than simply mating?

Prayer: Dear Lord Jesus Christ, at virtually every wedding we remember Your presence at the wedding at Cana where You did the first miracle of Your ministry. Help me always to keep my marriage relationship as You intended it, and through it become enriched. Amen.

Tetrakaidecahedron Skin

Job 10:8
"'Your hands have made me and fashioned me, an intricate unity; yet You would destroy me.'"

Dr. David Menton is a biologist who believes that God created the heavens and the Earth in six days only some six thousand years ago. As he likes to point out, he is a biologist who evolutionists say cannot exist — a biologist who rejects evolution.

But Dr. Menton's accomplishments make it hard for evolutionists to deny his right to be considered a top-notch scientist. Before completing his doctorate, he worked as a dermatology researcher at the Mayo Clinic. After earning his doctorate, he worked as a research scientist at the University of Iowa, where he made a major discovery. Dr. Menton is credited as the discoverer of the structure of skin cells. Biologists had known that skin cells were basically flat and arranged in layers. Dr. Menton, however, discovered that the cells themselves are 14-sided polygons called tetrakaidechedrons — sort of fun to say when you get the hang of it. He also discovered that these flat tetrakaidechedrons are stacked and interlocked like poker chips, one exactly on top of another. From this he learned that when skin cells reproduce too rapidly they do not stack as they should, resulting in skin problems like psoriasis. No, there can be no questioning that Dr. Menton is a qualified biologist.

We wouldn't want you to think that Dr. Menton is the only scientist, or even the only biologist, who believes in creation as taught in the Bible. Thousands of qualified scientists believe in creation, despite what some evolutionists say.

Prayer: Dear Father, I give thanks to You for all of the Bible-believing scientists who are working today. Protect them and give them boldness to witness Your truth. I also ask that You would give more of our Bible-believing young people the desire and strength to go into science professions. In Jesus' Name. Amen.

The Most Complex Structure in the Universe

Job 38:36
"Who has put wisdom in the mind? Or who has given understanding to the heart?"

Whether you know it or not, you are the proud owner of the most complex arrangement of matter in the universe. I am speaking of your brain. Even though it weighs only three pounds, the brain's structure is so complex that it defies all explanation except the explanation which includes a Creator.

The average brain has ten billion neurons. Each neuron has some 25,000 connections to other neurons, making the total number of connections in your brain something on the order of ten billion times 25,000. It has been estimated that the brain is able to hold one million times as much information as anyone could possibly learn in a lifetime. If true, this suggests that man was meant to live much longer than 70 or 80 years.

Thanks to the brain, you can listen to someone talk and begin to put your reply together for him while your brain is also directing a number of other projects, like your heart beat, breathing, and blood chemistry without your conscious attention. Your brain, through its extended sensory system, allows your finger to feel a vibration of less than eight-one-thousandths of an inch or see 10 million different colors.

The brain is so marvelously complex that even scientists who believe that the material is all that exists have admitted that the brain leads them to wonder if there is not more to reality. Some are now even willing to talk about a spiritual reality.

Prayer: Dear Heavenly Father, I thank You for the gift of my brain and all of the abilities it gives me. Help me to use my mind and all its powers to do Your will, and in this way love You and serve You with all my being. In Jesus' Name. Amen.

Doctors Use God's Invention

Job 10:10-12
"Did you not pour me out like milk, and curdle me like cheese, clothe me with skin and flesh, and knit me together with bones and sinews? You have granted me life and favor, and Your care has preserved my spirit."

Man has, for thousands of years, applied a huge variety of things to his skin to help injuries heal. By trial and error he has come up with ointments which keep injured skin moist so that it will not be hindered from healing naturally. Man has developed antibiotic ointments to prevent infection while skin heals naturally. But man has never come up with anything that actually increases the rate at which skin naturally heals.

That has now changed, although technically it was not man with all his knowledge and science who developed epidermal growth factor. Epidermal growth factor, made by the body, is one of the chemicals which God created to help keep the body in working order. Man is only now learning how to copy the chemical and use it. Tests have shown promise that epidermal growth factor can speed the rate at which wounds heal. Studies show that wounds on which epidermal growth factor is used heal about a day-and-a-half faster than normal. That might not seem like much, but when burn victims need many skin grafts, epidermal growth factor could mean the difference between life and death.

With all his knowledge of the chemistry of biology, one would think that man would be able to develop better healing agents than mindless evolutionary chance. Of course, the reason man's intelligence cannot easily outdo nature is because man is not trying to out-think mindless forces, but the Creator God. Epidermal growth factor is but another example of our Creator's practical wisdom!

Prayer: Dear Lord, as I look around Your creation, I see so many ways in which Your knowledge and wisdom is so obviously practical. Help me to understand that all of Your wisdom, including what You tell me in Scripture, is really practical and useful for life. Amen.

Do Christians Need Evolution?

1 Corinthians 15:21-22
"For since by man came death, by Man also came the resurrection of the dead. For as in Adam all die, even so in Christ all shall be made alive."

Many, if not most of the leaders of the creationist movement today were once evolutionists. Most of them believed that God created but that He used evolution to do it.

Those who were once convinced evolutionists now point out that there are some very serious objections to evolution. For example, the fossil record offers no record of the development of living things from simple to complex. If you are a theistic evolutionist you might not believe this, but check it out: representatives of every major phylum of animals living today are found in the very deepest layers of rocks containing multi-celled fossils. They appear suddenly without any evolutionary ancestors.

Further, the fossil record offers not one undisputed "missing link" between the various kinds of creatures. Dr. Colin Patterson, head of paleontology for the British Museum, who is not a creationist, admits in a well-publicized letter that there is not a single transitional form to link any kind of animal with another. Even the most famous of the so-called transitional forms have not stood the test of scientific scrutiny.

Even more important, the Bible clearly tells us that there was no death before Adam sinned. So whether you look at science or at the Bible, there is no reason for Christians to adopt evolutionary beliefs and every reason to believe the Bible's account of creation!

Prayer: Dear Father, the belief in evolution, even the belief that there was death before Adam, has done great damage. I pray for the work of those who are showing that it is not necessary to believe in evolution. Show me how I can help. In Jesus' Name. Amen.

Vacuum Cleaner Lungs

Genesis 2:7
"And the LORD God formed man of the dust of the ground, and breathed into his nostrils the breath of life; and man became a living being."

Man has always had to deal with air pollution. The air is filled with dust particles. And whether it is dust blown off the ground, smoke particles, or tiny particles of fabric or paper in the house, some of these particles find their way deep into our lungs.

In His wisdom, our Creator has provided us with tiny, but effective "vacuum cleaners" in our lungs which collect and remove the particles which make it into the lungs. These "vacuum cleaner" cells are actually a specific type of macrophage. They work in the deepest part of the lung, reaching out with tubular extensions which engulf a particle in much the same way that an amoeba eats. This "vacuum cleaner" essentially eats the particles it finds. Each one of these "vacuum cleaner" cells can clean an area forty times its own size.

Scientists say that they still do not fully understand whether certain particles, like poisonous heavy metals, affect the working of these cells. They suspect that certain particles may prevent these "vacuum cleaners" from effectively doing their necessary work.

Though science has known about macrophages for over a century, scientists are just beginning to learn how these cells actually clean the lungs. We can be very thankful that we have a Creator Who is not only clever enough to think of our need, but wise enough to know how to design a solution.

Prayer: Dear Father, I thank You that You have created so many mechanisms to help keep my body healthy. Help me to be as concerned about my health as You are so that I may enjoy the life You have given me and be in better condition to serve You. In Jesus' Name. Amen.

The Ten-Hearted Fertilizer Factory

Psalm 22:6
"But I am a worm, and no man; a reproach of men, and despised of the people."

The lowly earthworm is one of the most under-appreciated but important creatures on Earth. People probably don't appreciate the earthworm because it's not very pretty. Though evolutionists tell us that creatures like the earthworm were among the first land creatures to evolve, the earthworm is really not a simple creature at all. It has a well-developed nervous system and brain. It even has ten hearts!

Earthworms tunnel through the soil by literally eating the soil itself. As the worm draws nutrition from organic matter in the soil, its digestive system adds chemicals to the soil which are excellent natural fertilizers for plants. In one recent test, soil scientists held a competition between earthworms and the best fertilizers you can buy. The earthworms easily won the contest. The ground they fertilized had five times more nitrogen, twice as much calcium, two-and-a-half times more magnesium, and seven times more phosphorus than the best that modern science could produce. Since earthworms like lots of organic material in the soil, you can encourage earthworms in your garden by digging lawn clippings and most types of tree leaves into your soil this fall, rather than throwing them away. One note — some acid leaves like red oak are best left out of the garden.

So, while earthworms are not pretty, they are a wonderful invention of our Creator to enrich the soil and help feed billions.

Prayer: Dear Lord, some of Your best gifts to us come in humble and lowly form. Indeed, the world's wisdom leads it to scorn the Savior of the world and Your plan of salvation. Protect me from such false wisdom. In Jesus' Name. Amen.

Doubters Stand Back, We're Excavating the Bible!

Joshua 10:10-11
"So the LORD routed them before Israel, killed them with a great slaughter at Gibeon, chased them along the road that goes to Beth Horon, and struck them down as far as Azekah and Makkedah."

Over two thousand years ago, Bible critics were attacking the Bible as being full of errors and unscientific. Modern versions of Bible-doubting scholarship began late in the eighteenth century. By the time Darwin began to popularize his evolutionary ideas, many leading so-called Bible scholars had already declared much of the Bible to be nothing more than error-filled human writings.

At that time, biblical archaeology had done little to excavate the ancient cities which are talked about in the Bible. Yet scholars felt free to consider this lack of evidence as proof that people, cities, and even whole nations mentioned in the Bible never existed. It didn't seem to matter to them that they had never even bothered to look for evidence. But then biblical archaeology started making discoveries. And with each discovery, the critics' pronouncements about errors in the Bible began to fall. One example is the biblical record of Joshua's leadership of Israel in occupying the promised land. Joshua 10:10-11 reports how Joshua defeated Hazor, burning the city to the ground. Though this had been called into question, archaeologists found the ruins of Hazor. The ruins clearly showed destruction by an intense fire. The fire had been so hot that adobe had been baked lobster red, stones had been changed by the heat of the fire, and the ashes of the city were five feet deep!

Every claim that I have ever heard of that there is an error in the Bible has been based on lack of knowledge. Let's face it, no matter what the subject, including history, God has more knowledge than we do!

Prayer: Dear heavenly Father, I thank You for Your Word which has been preserved for our instruction today. Guide me by Your Holy Spirit to make good use of Your Word in my daily life. In Jesus' Name. Amen.

Hydraulic Plants and Animals

Job 15:8
"Have you heard the counsel of God? Do you limit wisdom to yourself?"

Hydraulics are part of our everyday lives. When you step on the brakes in your car, brake fluid, because it cannot be compressed, is forced from a cylinder into the braking mechanisms attached to your wheels. Hydraulic machines work because fluid cannot expand, but will flow easily to areas where there is less pressure.

But man was not the first to invent uses for this principle. The spider uses it all the time. You see, the spider's skeleton is on the outside of his body and his muscles are inside his skeleton. Since our muscles are outside our skeleton, we can use muscles both to bend our legs and stretch them out. But the spider's arrangement only allows his muscles to bend his leg. When he wants to straighten out his leg to take another spidery step, he pumps fluid into his leg and the leg joint hinges open once again.

The North American Dwarf mistletoe builds up hydraulic pressure equal to that found in a truck tire in order to catapult its seeds out to a distance of almost 50 feet at a speed close to 60 miles per hour. The squirting cucumber found in the Mediterranean area uses the same principle to propel its seeds up to forty feet.

Unless you think that the brake system in your car came together by chance and natural forces, it doesn't make any sense to believe that spiders and other plants and animals came about without a Designer and Creator.

Prayer: Dear Lord, I am constantly amazed at the technical excellence of Your creation. I ask that more scientists would also be amazed, and draw the natural conclusion that You are God. From this realization, help them to be open to hearing what You have done for them and all of us through Your Son, Jesus Christ. In His Name. Amen.

Was There Enough Room in the Ark?

Genesis 6:14-15
"'Make yourself an ark of gopherwood; make rooms in the ark, and cover it inside and outside with pitch. And this is how you shall make it: The length of the ark shall be three hundred cubits, its width fifty cubits, and its height thirty cubits.'"

One of the standard arguments used by atheists and religious liberals against the biblical account of the Flood says that Noah could not have fit all the species of animals on the Ark. After all, they argue, there are over a million species of animals, so the Ark would have been too small.

The first problem with this argument lies in the word "species." Species is a modern, human classification, but Genesis says that Noah was to take "kinds." The biblical description of kinds leads us to believe that in many, if not most cases, the word kind is a broader term than species. A kind could include many species. For example, the dog kind has many different species, including the domestic dog, wolves and foxes. Many of these species can still interbreed with each other showing their close relationship. This drastically reduces the number of animals on the Ark to a more reasonable figure. That figure is further reduced when we realize that one-third of all species of animals are found in the sea and have no need to be saved from a flood.

But let's not forget that the Ark was the largest ship ever built by man until this century. It had a capacity of 1,396,000 cubic feet — about equal to a train made up of 522 railroad stock cars!

So when the facts are all considered, even a generous estimation of the space needed on the Ark tells us that there was room not only for Noah and his family and all the animals, but also for plenty of food.

Prayer: Dear Father, I ask that You would put to confusion all attempts to deny Your truth. Make Your people, including me, well-armed with Your Word and how it applies to all parts of life. In Jesus' Name. Amen.

Precise Blood

Genesis 9:4

"'But you shall not eat flesh with its life, that is, its blood.'"

Most people are aware of how important their circulatory system is to life. But many of us don't fully appreciate the almost miraculous efforts needed to keep us alive every second.

Your heart pumps about 100,000 times every day. That means that your heart pumps the equivalent of 10 tons of blood every day — 80 million gallons in a lifetime. Your circulatory system brings that blood to every cell in your body through a capillary network which is so large that the combined capillaries of only four people, stretched end to end, would reach from the Earth to the moon!

But your circulatory system involves quality as well as quantity. It is crucial to life that the chemistry of your blood be within very precise limits. Your body uses various chemicals, your breathing and kidneys, among other organs, to keep this balance. Take the example of how precisely your body controls the acid level of your blood. It keeps your blood acid level in control so that the difference between the highest allowed level and the lowest is only one-one-hundred-millionth of a gram!

Our circulation system is one of many cases where "almost" is not good enough. Minor changes in any one of a thousand factors could be fatal. So if mindless evolution created us, it had to make the whole system perfect the first time. There is no room for millions of years or for a poorer circulation system to be improved by mutations. Here science clearly supports Scripture in saying that we were created in finished form, not developed by trial and error.

Prayer: Dear Lord Jesus Christ, You Who created blood and everything else took our human form and nature and shed Your blood on the cross for my salvation. Take my life and let it be one of continuous thanks to You. Amen

Could You Spend Winter Outside?

Psalm 104:27-28
"These all wait for You, that You may give them their food in due season. What You give them they gather in; You open Your hand, they are filled with good."

If you live in the northern hemisphere, it is very likely that the approaching winter has made itself known in your area already. With that in mind, how would you like to move outside right now and stay there, never coming in until next May?

Human beings were not made for living outside like animals, despite what evolutionists say about our past. We need the protection of shelter, especially during the winter. But since animals cannot build houses, the Creator has provided them with other ways of surviving winter. Unlike humans, animals slow their metabolic rate in the winter so they need less food. Since there is generally less food available in the winter and the weather is often too poor to find food, the wisdom of this design is easy to see. In addition, most animals are able to store fat reserves to help carry them through the winter. The amount of fat they store and the amount that their metabolism slows are carefully coordinated, and are often very different, even between males and females of the same species.

If human beings had evolved from animals, we would expect a much greater similarity between how animals and humans cope with winter. But the fact is, we could not possibly survive under the same conditions in which millions of creatures not only survive but even continue their reproduction through the winter months. This is a clear but silent statement from our Creator that we are not animals, nor have we ever been!

Prayer: Dear Father, You care and provide for all Your creatures. Grant us seasonable weather this winter for the good of the earth and all living things, and for the productivity of the soil next growing season. In Jesus' Name. Amen.

Is the Shark a Living Fossil?

Job 12:7-8
"'But now ask the beasts, and they will teach you; and the birds of the air, and they will tell you; or speak to the earth, and it will teach you; and the fish of the sea will explain to you.'"

There are over 350 species of shark. These remarkable creatures range from about the size of a shrimp to the whale shark which can be over 60 feet long. Most people view sharks as an ancient form of primitive fish, often because they are popularly presented to the public as a so-called "living fossil."

The truth is, sharks are not primitive or simple. In fact, scientists who classify living things are uncomfortable classifying sharks as fish, preferring instead to call them "fish-like vertebrates." Unlike all other fish, the shark has no skeleton. In fact, the shark's skin serves as the anchor for its muscles in much the same way that your bones do. Unlike fish, sharks have no gill coverings and their fins are structured in a completely different way than fish fins. Most familiar sharks even have a placenta very much like mammals'!

Sharks are also much more complex than fish. They have sophisticated sense organs and their brains are much larger compared to body weight, than most fish. Unlike fish, sharks can learn a route through a maze as well as laboratory rabbits. Sharks are also socially complex and they communicate with each other in a number of ways.

Amidst the many kinds of creatures God has created using His unlimited imagination, it appears that He followed some similar themes in each kind. This offers us a better principle than does evolution for classifying living things, which arranges them from simple to complex.

Prayer: Dear Father, I stand in awe at the ways in which Your unbridled creative imagination is expressed in the living world. Help me to be a better witness of this fact to the world around me, which is taken with the imaginations of its own heart. In Jesus' Name. Amen.

Not So Bird-Brained

Genesis 1:21
"So God created great sea creatures and every living thing that moves, with which the waters abounded, according to their kind, and every winged bird according to its kind. And God saw that it was good."

All of us have admired the beautiful feathers worn by birds. But while such beauty was a gift of the Creator to birds, it takes a lot of work and special provision for birds to keep their feathers beautiful. And the Lord has provided birds with a marvelously complex system to keep their feathers healthy and beautiful.

We have all seen birds preening. While the preening bird may look a bit vain, it is really going about a very serious business. Birds of the heron family accumulate a coating of slime on their feathers. To clean this condition they have three patches of feathers that break down into a powder that works like talcum powder. These birds apply the powder to their feathers. After the powder has absorbed the slime, it is combed out of the feathers using a special toe that is shaped like a comb. Many birds also oil their feathers after cleaning them using oil produced by an oil gland. This waterproofs their feathers.

Many birds even use a pesticide to get rid of parasites. There are hundreds of species which will sit near an ant hill. Believing they must protect their nest, the ants crawl onto the bird and release formic acid, which drives off mites and most other pests.

God has not only provided birds with beautiful feathers, He has also given them everything they require to keep their feathers beautiful and healthy. How happy we would be if we could simply lay every one of our cares and needs, no matter how small, on Him as He invites us to do.

Prayer: Dear Father in heaven, cleanse my mind and heart of the thought that You have only provided for my greatest needs, like forgiveness in Jesus Christ. Help me to trust You more completely for everything. In Jesus' Name. Amen.

The Weaverbird

Genesis 1:28
"Then God blessed them, and God said to them, 'Be fruitful and multiply; fill the Earth and subdue it; have dominion over the fish of the sea, over the birds of the air, and over every living thing that moves on the Earth.'"

Intelligence — as much as is needed and no more — is widely found throughout the creation. So intelligence alone is not what makes man special.

Consider the weaverbird. The nest of the weaverbird is woven of strips of fiber and grass. Using beak and feet, the male uses loops and knots to weave its hanging nest. Then the nest must be inspected by a prospective mate. If she doesn't like the nest's construction, she turns down the hopeful male. The male must then tear down his work and start over. Some males have been observed constructing and tearing down their nests two dozen times before a prospective mate is found who is satisfied with his work. Some weaverbirds actually build huge cities of nests protected by a woven roof. One roof over a weaverbird city was 15 feet across!

While man's intelligence spans much more than animal intelligence, what sets us apart from animals is the fact that our Creator made us to have a special relationship with Him. And even when we placed our will above His Word, He still loved us enough to pay a price to restore us to Himself. Jesus Christ lived in perfect obedience to God for us, then suffered the penalty of our disobedience against God. In His Resurrection from the dead, all who embrace Christ in faith have the promise of being made new creations once again — beginning right here in this life! That's the wide gulf between man and animal!

Prayer: Dear Father, I thank You that You have given me being and life, and that when I and mankind were lost in sin, You still sought me out with Your gospel. Help me to truly live as Your new creation in Christ. For His Sake. Amen.

Chocolate Chili?

Genesis 1:29
"And God said, 'See, I have given you every herb that yields seed which is on the face of all the earth, and every tree whose fruit yields seed; to you it shall be for food.'"

It has been written against, outlawed, and is the original "food of the gods." Modern science is finding that chocolate seems to offer some positive health benefits.

When Cortéz and his soldiers came to the New World, Montezuma offered them his people's favorite hot chocolate. But Montezuma's people didn't add sugar to the chocolate. Instead, they added ground up hot peppers. This was *real* hot chocolate! The Spanish thought it was terrible stuff. Luckily for chocolate lovers they did take some of the beans back to Europe. There it dawned on someone to leave out the peppers and add sugar instead. By the eighteenth-century, chocolate houses had become so popular that brewers of alcoholic beverages demanded legal restrictions against chocolate because it was ruining their business. Today, Americans eat an average of 10 pounds of chocolate per person per year.

One study, which has been disputed, suggests that chocolate may help lower cholesterol. Chocolate also contains a mild anti-depressant as well as theobromine, a unique substance which perks you up and calms you down at the same time. And contrary to popular thought, chocolate is not addicting.

While chocolate may seem a trivial subject, it has given those who like it a lot of enjoyment. It is one of those gifts God has given us that makes life a little more enjoyable — but is totally unnecessary. And this offers us a lesson about His gracious thoughts toward man.

Prayer: Dear Lord Jesus Christ, You made all that is. I thank You that You have made some things which we don't require, but which cause us to rejoice in Your generosity. Help me never to misuse Your good gifts. Amen.

Smart Sharks

Psalm 148:7
"Praise the LORD from the earth, You great sea creatures and all the depths. . . ."

Sharks have larger and more complex brains than fish. In fact, in learning tests they score about as well as rabbits. That won't get them into college, but it does show that they are not some prehistoric left-over of evolution.

Over two-thirds of the shark's brain is used to sense smells in water. In addition, the shark has sensors just beneath its skin, running the length of its body which sense low-frequency vibrations. Beneath their heads, sharks have pits which are equipped to detect the electrical fields given off by other living things, and which may also help them navigate.

Perhaps the most amazing feature of shark intelligence is that they are socially complex creatures that communicate with each other through a language of body signals. Researchers have so far learned what seven of these signals mean. For example, as sharks circle in a school, the larger and more dominant sharks move toward the center one by one, using a number of body signals to direct other less dominant sharks toward the outside of the circle. Male hammerheads show submission to other hammerheads by shaking their heads.

Because intelligence and ability are gifts of the Creator, they are often found in abundance in creatures which evolution says are simple or less-evolved creatures. This fact tells us that the Bible's description of the creation in which we live is more accurate than conflicting theories which man dreams up.

Prayer: Dear Lord, You are truly Lord of all the creation, having made nothing of which You are ashamed. Help me to have this same attitude toward all of Your creation. Amen.

Created Male and Female

Genesis 6:19
"'And of every living thing of all flesh you shall bring two of every sort into the ark, to keep them alive with you; they shall be male and female.'"

One of our *Creation Moments* listeners has written to ask how evolutionists explain the development of male and female. The question is, if a mutation produced the first male, it isn't likely that another mutation would have produced the first female at the same time and in the same neighborhood. He further pointed out that studies now show that with the current genetic errors we all carry within us, one male and one female would not be enough to establish a new population of male and female creatures.

The question is important. Evolutionists have admitted that their theory does not have a satisfactory explanation for how male and female could have developed. One evolutionist even noted that because evolutionists have no credible explanation, most textbooks simply ignore the question as obvious as it is.

But evolutionists also point out that the problems in explaining the origins of male and female are even greater than those I've already mentioned. They now admit that their studies show that if creatures progress by evolution, creatures who reproduce sexually would be at a disadvantage. Even worse, sexual reproduction is, as one evolutionist put it, designed to weed out the very genetic variations which supposedly cause evolution!

It is difficult to understand, then, how anyone could say that evolution offers a better explanation of life than the Bible does.

Prayer: Dear Father, I thank You that in Your wisdom You have made us male and female, and that You have done so in a way which confounds man's rebellious wisdom. Help this fact be a witness to Your glory. In Jesus' Name. Amen.

God's Chemistry Again Outpaces Man's

Psalm 139:14
"I will praise You, for I am fearfully and wonderfully made; Marvelous are Your works, And that my soul knows very well."

One of the greatest questions in biology asks how a single fertilized cell divides into many different cells — some become liver cells, skin cells, brain cells and bone cells. This is the ultimate way of asking the question, "Where do babies come from?"

Scientists are now studying a chemical produced in the body called human growth factor to see if it plays a part in the development of all the billions of specialized cells in our bodies from one cell. But human growth factor is only one of about 30 known chemicals in the body which stimulate and guide the development of new cells. Scientists studying these substances describe their power as almost mystical. Another powerful chemical, called GM-CSF for short, actually stimulates the development of white blood cells by stimulating the development of bone marrow. This drug has saved otherwise doomed patients who have lost virtually all of their bone marrow because of exposure to radiation. Nerve growth factor is being studied as a possible treatment for Alzheimer's disease and for use in re-attaching limbs.

Do you notice where all of these incredible chemicals come from? They don't come from laboratories staffed with modern science's best minds. They are produced by our bodies. Science can only stand in awe and try to learn about the wisdom with which these substances are constructed. And that's one of the best arguments that we were created by a wise and powerful Creator, not by mindless chemical accidents!

Prayer: Dear Lord, I am truly fearfully and wonderfully made. I ask that you would guide medical researchers so that they may learn how to use the substances You have created for the earthly betterment of our condition and that You would be glorified through this. Amen.

Built-In Error Correction

Proverbs 3:11-12
"My son, do not despise the chastening of the LORD, nor detest His correction; for whom the LORD loves He corrects, just as a father the son in whom he delights."

It seems like science fiction to suggest that if someone typed your name into a computer, misspelling your name, the computer would find and correct it.

But there is an even more sophisticated information storage and transmission system in each of your cells — your genetic code. Not only does your genetic code store far more information in a microscopic space than our largest computers can, it has a built-in error correction system. Scientists have discovered a number of key enzymes within the cell that have just one job: find and correct errors in the genetic code. Those errors can creep into the code because of radiation, certain chemicals, or other outside forces. The enzymes faithfully correct any errors, preventing mutations — the same mutations which scientists thought could cause evolution. One British scientist who studied these enzymes stated that scienctists have no explanation for how this could have evolved naturally. Furthermore, they don't know how life could have gone on without this genetic proof-reading and correction.

Only our Creator could have been wise enough to design an information system which can correct its own errors. Even we human beings have not figured out how to do this with our much simpler computers. Certainly blind chance and natural law could not do what we cannot figure out how to do!

Prayer: Dear Father, just as You have designed a chemical system to remedy errors in our genetics, Your plan of salvation by grace through faith in Jesus Christ's atoning work gives the full remedy for sin, including a new life. Grant me Your Holy Spirit so that I may be enabled to better live that new life you give me. In Jesus' Name. Amen.

How to Make a "Bananatrode"

Psalm 147:5

"Great is our Lord, and mighty in power; His understanding is infinite."

On his way to the lab, a scientist stops by the grocery store to pick up a banana. At the meat market, he asks for a piece of antenna from a blue crab. In the lab, he hooks up the banana and the antenna, one at a time, to an electrode which is wired into his computer. Having completed his day's work in record time, he begins looking for a whisker from a catfish on the way home.

What sounds like the plot of some new television situation comedy is really the plot of a story in which supposedly simple, less evolved creatures produce faster and more sensitive chemical sensors than our best scientists can. The "bananatrode" detects a chemical produced by the human body called dopamine. The crab antenna detects tiny traces of amino acids in salt water. And a catfish whisker can detect equally small amounts of amino acids in fresh water. Crabs, you see, find their lunch by sensing and tracing down amino acids in the water. These sensors work much faster than other methods of chemical analysis, since crabs can't send water samples out to a lab for analysis. They must identify and trace down clues to lunch on the spot. These natural sensors are being studied for human medical uses. One such application is a glucose sensor for an artificial pancreas.

It certainly was wise of our Creator to give some of His creatures these wonderful abilities so that they could make their livings and to do so in a way in which man can also use them for his benefit.

Prayer: Dear Father in heaven, You have filled Your creation with wonderful miracles of design so that men would seek You Who made them all. Preserve me from the pagan sin of putting the creation over You, the Creator. In Jesus' Name. Amen.

Occupational Studies Among Honeybees

Psalm 19:9-10
"The fear of the LORD is clean, enduring forever; the judgments of the LORD are true and righteous altogether. More to be desired are they than gold, yea, than much fine gold; sweeter also than honey and the honeycomb."

How would you like the job of gluing tiny, numbered, color-coded tags to the backs of 7,000 living honeybees? Well, someone had to do that in a recent study of how honeybees select their specialized jobs within the hive.

What scientists were hoping to learn was whether honeybees select their occupations based on what their fathers did, or for some other reason. Typically, the queen bee mates with over a dozen males before settling down to a year or two of continuous egg-laying. In one study, the queen was allowed to mate only with a "guard bee," and an "undertaker bee," whose job is to dispose of dead bees in the hive.

What scientists learned was that a young bee is likely to take up its father's profession 80 percent of the time. This has been interpreted to mean that much of the complex social structure of the bee hive is built into the genetic structure of the bee. But the study also left evolutionary scientists with a mystery. There is a wide genetic variation within the bee hive, so if the whole arrangement would have evolved, they would not expect bees to have such a cooperative and complex society.

The obvious answer is, of course, that when something is clearly so intricately designed there must be a Designer. Evolutionary scientists realize this. Some have even admitted that their main objection to believing in creation is that they would have to accept the fact that there is a Creator.

Prayer: Dear Lord, I pray that I shall never desire to evade You, either by denying my need for Your forgiveness, by denying my sin, or by denying Your work of creation. For Jesus' sake. Amen.

Mouse Doctor Finds Wisdom in Fungus

Daniel 2:22
"He reveals deep and secret things; He knows what is in the darkness, and light dwells with Him."

Fungus grows in the dirt virtually all over the world. After millions of dollars of research on the fungus, looking for a usable drug, medical researchers were ordered to stop their research. But one scientist disobeyed orders. And the result is the most important miracle drug since the discovery of antibiotics!

When Jean-Francois Borel, who describes himself as a "mouse doctor," refused to give up research on a white fungus first identified in soil samples from above the Arctic Circle, organ transplant surgery had virtually stopped. Organs could be transplanted, but the body's rejection of the organ killed over half of all transplant patients within a year of the transplant. Dr. Borel's continuing research led to the discovery of cyclosporine, a chemical which prevents the immune system from rejecting transplanted organs. There are now over 150,000 people living with transplants because of Dr. Borel's discovery. Yet scientists are still not sure how cyclosporine works.

Fungus might seem like an odd place to look for wonder drugs. But drug companies have learned that microorganisms in soil produce a wide range of medically interesting chemicals. Each year they test some 30,000 samples to see whether a miracle drug might be hidden in one of them.

So, although many scientists say they believe there is no purpose or mind behind nature, they regularly look for purpose in the creation.

Prayer: Dear Lord, as man searches out the benefits offered by the things You have created, do not allow us to lose sight of our greatest need, a relationship with You. This cannot be found in nature, but only in Your Son, Jesus Christ. In His Name. Amen.

Designed to Eat

Job 36:5
"'Behold, God is mighty, but despises no one; He is mighty in strength of understanding."

The shark seems to be designed for one purpose and one purpose only — to eat.

Up to two-thirds of the shark's brain is devoted to its excellent sense of smell which enables it to find food. In its mouth are many rows of sharp teeth and when one set wears out or has some broken teeth, another set behind it moves forward as if on a conveyor belt. The shark's stomach has an unusual spiral valve so that more nutrients can be absorbed in less space — even though the shark's stomach is large. The shark's liver can account for as much as a quarter of the its entire body weight. It only seems fitting that a creature that starts life eating its own brothers and sisters while still in its mother's womb is literally covered with thousands, if not millions of teeth. That's right, sharks don't have scales or a leathery hide. They are literally covered with teeth. Called dermal denticles these are true teeth which are covered with dentine and have a central pulp canal. Meshed closely together as they are, they provide the shark with a strong armor.

It is also true that sharks seem willing to eat anything. The contents of one large shark's stomach included three overcoats, a raincoat, a driver's license, one cow's hoof, deer antlers, twelve lobsters, and a chicken coop with feathers and bones inside!

The awesome shark stands as another example of God's limitless imagination and ability to create whatever He can think of. Scripture often refers to this as God's remarkable understanding.

Prayer: Dear Father, there are indeed many dangers to us in the creation; they often distract me from seeing that You are in charge and can do anything. Renew my vision and my faith. In Jesus' Name. Amen.

God Gives the Rat a Secret

Job 38:4
"'Where were you when I laid the foundations of the earth? Tell Me, if you have understanding.'"

A new revolutionary design in metal-cutting blades is changing the metal-working industry. These blades do not cut like a knife. Instead, the blade is fixed in one position while the metal to be shaped spins on a lathe. This new blade stays sharp six times longer than the old blade design. And where titanium used to make the old-style cutting blades dull almost the second they were used, the new blades last up to 30 minutes. This wonderful new blade was developed by two engineers who got the original idea from a rat.

You see, rats' teeth always have a sharp edge. Engineers learned that this is because the teeth are hard on one side and soft on the other. So as they wear, they always keep a sharp edge. Applied to metal cutting technology, this principle has lowered costs and enabled productivity to be increased, since cutting blades don't have to be changed as often.

Beginning in Job 39, God humbles Job with several chapters of questions about the wise designs found in the animal kingdom. God's message is, "Job, if you think you're so smart, Who created all this with such wisdom?"

Even the teeth of the rat are created with such wisdom, that once we learned the principle, our technology was greatly improved. In light of this it doesn't seem at all scientific to say that there is no Creator!

Prayer: Dear Father, help me not to think that I am wiser or smarter than I really am. Help me to seek Your wisdom and knowledge in Your Word, which is as high as heaven and as real as the creation. In Jesus' Name. Amen.

Delicate, Precise Designs

Matthew 6:28
"'So why do you worry about clothing? Consider the lilies of the field, how they grow: they neither toil nor spin. . . .'"

The way in which some creatures respond directly to the habits of totally different creatures upon which they depend, defies the idea of a chance relationship developing between the two creatures.

Take chicory for example. Chicory has a beautiful flower stem which it maintains from June through October. But it spaces out the opening of its flowers through these months on a precisely timed schedule which is in perfect harmony with its pollinators. Early in the season, chicory flowers nearest the bottom of the flower stalk open. As the summer moves through to October and the lower flowers are pollinated, they close while the flowering sequence moves up the stalk. In this way, chicory is not as dependent on the weather as plants which flower for only a short period. Chicory flowers are usually open only in the cool morning hours when its pollinators are about. Then they close as the sun rises in the sky to protect the delicate flowers from sun and heat. On cloudy days the flowers are open longer. Like many flowers, chicory's flowering cycle is closely coordinated with the creatures which pollinate it. Other flowering plants open their flowers only at night and tend to have white flowers so that they are easily seen in the dark because they are pollinated by night flying moths.

Such coordinated inter-relationships between widely varying creatures could not have developed through accidental situations. They provide more evidence that the wise Creator designed these and all creatures within their relationships from the beginning!

Prayer: Dear Lord, in Your design of the creation You have seen to the needs of all living things. Take from me a mind which thinks that I must do for myself what You will do for me, and replace it with faith in Your promises. Amen.

Plants' Natural Pest Control

Isaiah 40:29
"He gives power to the weak, and to those who have no might He increases strength."

Most of us know that a plant under stress is more likely to have problems with insects. If a plant is dry, for example, we water it, hoping it will regain its health and stand strong against insect attacks.

But in the wild, no one shows up with a watering can when it gets a little dry. Ironically, most of us are trying to get our own plants to grow as well as they do in the wild. Now science may have discovered a provision the Creator has built into plants to protect them from slight water stress.

Scientists have found that severe water stress is bad for plants, but that mild water stress actually aids plants in fending off insects. First of all, plant leaves under water stress become tougher to chew or penetrate and contain less nourishment. In addition, the stressed leaves begin producing mild insecticides when they are short of water. Some plants also start producing waxes and other substances to protect themselves under these conditions. Once water stress is relieved, plants can, overnight, return to normal operation.

The Creator has wisely provided for both plants and insects by means of this balance. When plants have plenty of water and can afford to provide food for insects, insects may lunch away, undeterred. When a plant has barely enough for itself, it is given the ability to fend off insects who would feed on it so that they can find other sources of nourishment.

Prayer: Lord, sometimes in our sinful world, we are tempted to accept the idea that life really is survival of the fittest. As a result, our love grows cold and our tolerance of what we should tolerate is short. Forgive and strengthen us, for Jesus' sake. Amen.

Evolutionary Prediction Fails

Deuteronomy 18:22
"'... when a prophet speaks in the name of the LORD, if the thing does not happen or come to pass, that is the thing which the LORD has not spoken; the prophet has spoken it presumptuously; you shall not be afraid of him.'"

If man is the result of billions of years of evolution from the simplest creatures, evolution predicts that man should have most if not all of the abilities of other living things.

According to evolution, our ape-like ancestors had opposing toes, just as we have opposing thumbs. If we had opposing toes, we could pick things up without bending over. The female chimp can pull 1,260 pounds with one arm. Such strength would certainly be helpful to man. In fact, the hero shrew of Uganda which is only six inches long, can support the weight of a 160 pound man on its back!

Owls can see 100 times better than we can at night. The golden eagle can see a rabbit two miles away. Fleas can jump 130 times their own height enduring a force of 200 gravities. If we could do that, any one of us could jump from the sidewalk to the top of the World Trade Center towers in New York! Even the lowly snail can pull up to 200 times its own weight, and lift 10 times its weight. The trilobite, supposedly a primitive creature which, by inflated evolutionary years, is said to have been extinct for 300 million years, had the most sophisticated eye lenses ever seen in any creature!

Since all of these abilities would be greatly beneficial for us, why didn't evolution let us keep or develop these abilities as we evolved? The answer is simple. We have not evolved. Rather, we have been made by a Creator Who made us for the purpose of a relationship with Himself through His Son, Jesus Christ.

Prayer: Dear heavenly Father, I thank You that while I cannot leap tall buildings or lift them, You have made me for the higher purpose of a relationship with You. Increase my trust in the atoning work of Your Son for me so that I may grow closer to You. In His Name. Amen.

He Walks on Water

Isaiah 55:8
"'For My thoughts are not your thoughts, nor are your ways My ways,' says the LORD."

Whenever water comes in contact with air, one of the most delicate structures in the universe is created. That structure is called surface tension. Surface tension is created because water molecules touching the air move just a little bit closer to one another, forming a delicate skin on the water.

Surface tension is so delicate that not even most small insects can walk on it. But the water strider is an insect which has been given some very special equipment so that it can literally walk on water. Each of the six legs of the strider ends in a foot, which looks rather like a snowshoe. Actually, each water strider foot is covered with many sensitive hairs which allow its weight to be spread out enough that the surface tension of the water will support his weight. But if it were not for another provision, those hairs would soon become water logged, and the strider would sink. Each foot is equipped with an oil gland which lubricates the hairs so that water cannot penetrate them. In addition, the hairs pick up vibrations in the water which give the water strider information about other nearby creatures. The strider's body is also covered with these hairs, providing for flotation when it dives for food.

Only the Creator of all things could have known about surface tension and how to make a creature which can literally walk on water. Doesn't He deserve our worship and praise, especially since He did not even spare the life of His only Son to save us when we were lost to Him?

Prayer: Dear Father in heaven, You can do all things and with You nothing in or out of this world is impossible. Help me to remember that Your thoughts are far above ours, and that with Your unlimited creativity, power, and wisdom, You also love me and will take care of me, for my Savior's sake. In His Name. Amen.

Why Birds Don't Need Sox

Job 12:7,9
"'But now ask the beasts, and they will teach you; and the birds of the air, and they will tell you; Who among all these does not know that the hand of the LORD has done this?'"

There stand some ducks out on the ice, watching their mates paddle around in the near-freezing water. They will spend much of the day in one spot, the temperature well below freezing, and yet the cold doesn't bother them. Perhaps on shore there are some sparrows hopping around in the snow. And one wonders, why don't birds need sox?

If you or I ran around outside with bare feet and legs when the temperature was below freezing, it would not be very long before we had a good case of frostbite. We could even lose enough body heat to threaten our lives.

Birds have a network of arteries that all blood going into their legs must enter. These arteries are interwoven with the veins returning from the foot. As the blood going to the foot enters this structure, called a "wonder net," its temperature is 106 degrees F. The blood returning from the foot at the same time is only 37 degrees F. The warm blood, passing through the net, reheats the cold blood coming from the foot before it enters the body. The result is that the bird loses very little body heat, and the blood going into the foot never becomes dangerously cold in normal winter weather.

The wisdom of this design is obvious. Could birds have survived having their feet frozen off before they learned to evolve this special arrangement? Our only conclusion can be that there is a Creator Who cares for all His creation, including you and me!

Prayer: Dear Father, You have spared nothing for our good. I especially thank You that in Your wisdom You designed a plan of salvation for us, even though it cost You the life of Your only Son to purchase me. Help me never to treat Your wise plan of salvation lightly. In Jesus' Name. Amen.

God's Unusual Fig Arrangment

Habakkuk 3:17-18
"Though the fig tree may not blossom, nor fruit be on the vines; though the labor of the olive may fail, and the fields yield no food; though the flock be cut off from the fold, and there be no herd in the stalls; Yet I will rejoice in the LORD, I will joy in the God of my salvation."

Completely unrelated creatures who depend completely upon one another to live, having no way of living without one another, pose a serious challenge to those who say that living things simply evolved.

One startling example is the relationship between the fig wasp and the fig tree. Male figs, called caprifigs, are not for eating. They produce the pollen which grows the sweet, juicy figs of the female fig, called the calimyrna or Smyrna fig. But the flower parts of both the male and female figs are inside the figs. There is no chance that the wind could spread pollen from the male to the female fig. And there is no other creature besides the fig wasp who spreads the pollen.

Herein lies an incredible story. The fig wasp hatches from eggs inside the male fig. As the wasp hatches, it has no other thought in mind but to lay its eggs inside another male fig, for by nightfall it will be dead of old age. The wasp emerges from the male fig, covered with pollen. It is programmed to search for a good place to lay its eggs in a number of female figs, thus pollinating them, before laying its eggs inside a male fig.

The fig wasp lays its eggs nowhere but within the male fig. And the fig cannot be pollinated by any other method. Clearly, both the fig wasp and the fig tree were created for this particular relationship.

Prayer: Dear Lord, just as You created the fig wasp and the fig tree for this close relationship, You have created me for a relationship with You. There is no peace for me outside of a relationship with You. Help me to grow closer to You through the instruction of Your Word. Amen.

White Death

1 Corinthians 15:54
"So when this corruptible has put on incorruption, and this mortal has put on immortality, then shall be brought to pass the saying that is written: 'Death is swallowed up in victory.'"

When the subject comes up, most people say they are glad that we don't have to share the Earth with the great dinosaurs.

But we may, in fact, be living with a more dangerous man-eater than dinosaurs ever were. The Australians call it "white death" and we know it as the great white shark. The great white shark has been called the last free predator of man. Its favorite food is not necessarily man, but any mammal it finds in the water. This dangerous creature has been recorded at up to 21 feet in length and over 7,000 pounds. Surprisingly, it is not strictly cold-blooded. It does maintain a body temperature well above the surrounding water temperature. Great white sharks attack differently than other sharks. Instead of swimming around its prey and poking it, the great white shark spots its prey and darts straight toward it. As it reaches its prey, it rolls its eyes backward, opens its mouth, and takes a bite. The attack can last less than a second. Then it moves away and waits for its victim to bleed to death. This last habit has saved more than one diver who was rescued by companions while a great white shark was waiting for him to die.

God has given all of us a will to live and a sense that death is our enemy. But Christians also know that even death has been conquered by our Lord and Savior, Jesus Christ. He rose victorious over death so that all who come to Him may have that victory too. In Christ, we have been saved from the jaws of death for a new life!

Prayer: Dear Lord, I thank You that You endured the consequences of my sin and the pain of death in my place. Because of what You have done for me, make me more willing to tell others about it so that they, too, may be snatched from the jaws of eternal death. Amen.

Not a Blank Slate

John 1:1
"In the beginning was the Word, and the Word was with God, and the Word was God."

It has become popular in many circles today to think that human beings come into the world as a blank slate; in a condition somewhat like very primitive early human beings. This thinking has led to the idea that language, for example, is no more than simply sounds without inherent value. Those with a more biblical view of man realize that Scripture teaches that language is a God-given universal to man. New research is supporting the biblical view.

Researchers at the University of Washington in Seattle recently released their study on how six-month-old infants respond to certain sounds. The infants were trained to look for a reward, like the sight of a mechanical bear beating on a drum, when they heard vowel or consonant sounds that are part of human speech and fitting into the category of the previous sound. Researchers discovered that even infants recognize and categorize basic human vowel and consonant sounds, as well as their variations. Other studies have shown that people in a broad spectrum of cultures agree on what basic colors look like, for example, which hue best represents "red." Researchers are continuing to test their new theory that speech sound categories may be built into human beings.

The theory that a basic idea of what language is and is not, is built into us has no evolutionary explanation. But it certainly supports the Bible's claim that we were created by the second Person of the Trinity, Who Himself is referred to as the "Word."

Prayer: Dear Lord, You are the Word through Whom I and everything else was created, and through Whom I am saved from my sin. Let the written Word, the Holy Bible, be my guide and instructor, that I may know You better and love You more. Amen.

A Few Questions for Evolutionists

Job 5:13
"He catches the wise in their own craftiness, and the counsel of the cunning comes quickly upon them."

Why do giraffes have long necks or kangaroos have pouches? Evolutionists answer that natural selection has favored the development of certain characteristics while discouraging and finally eliminating other features. But if this is what happened, we, who believe in creation have a few questions.

Giraffes have long necks, say evolutionists, because conditions favored the development of long-necked creatures which could feed on higher parts of the tree. Well then, we ask, how did the sheep that live side by side with giraffes manage to get by? The horse, according to evolutionary explanations, has crowned teeth in order to survive in its environment. And yet the cow, with its uncrowned teeth, survive quite well in the same environment.

Some evolutionists say that plants developed berries so that their seeds, inside the berries, would be carried far and wide by hungry birds, thus ensuring the plants' survival. Why then did some plants develop poisonous berries? And if the maternal instinct evolved to preserve the next generation, why do creatures like the stickleback fish, sea horse, midwife toad, and Tilapia, to name a few, leave total care of the young to the male?

The truth is that natural selection does not offer a clear and consistent explanation for the living world. The diversity of the created world does not bear witness to evolutionary principles, but to the artistry of our Creator God.

Prayer: Dear Father, You confound those who are wise in their own hearts and give wisdom and clear vision to those Whom You have made pure through the blood of Christ. Let the wisdom and vision I seek be that which You provide. In Jesus' Name. Amen.

Was Noah's Ark Seen in 1989?

Genesis 8:4
"Then the ark rested in the seventh month, the seventeenth day of the month, on the mountains of Ararat."

As they rose toward the 15,000-foot plateau on the west side of Ararat, they spotted a familiar looking dark form. It looked almost like the drawing of Noah's Ark developed from the descriptions provided by others who had seen what they thought was the Ark in this area. The date was September 15, 1989.

Chuck Aaron, of Orlando (Florida) Helicopter Service and Bob Garbe, President of the Creation Research Science Education Foundation and a member of the Bible-Science Association's board, estimated that the visible portion of the dark object was about 60 feet wide, 30 feet high, with about 225 feet of it extending out of the ice and snow. The next morning they returned to the site and shot more video and photographs. With a storm moving in, working time was limited.

The following year other researchers returned to the site in a helicopter. While they could not land, they were able to get much closer to the site to get better pictures. From their vantage point, they concluded that the object appeared to be an unusual rock formation — a typical story in the search for the Ark.

It's a good thing that our faith in the truth of God's Word does not depend on physical evidence. Many creationists feel that even if the Ark were discovered and moved to Times Square in New York, unbelievers would eventually concoct their own explanation to discredit the story of God's judgment in the Flood.

Prayer: Dear Heavenly Father, You sent the Flood as a judgment upon a willfully disobedient generation. Our generation, too, wishes to deny their moral accountability by denying that You acted as judge of the whole world in the Flood. Help me to take comfort in the fact that, like Noah, I shall be saved by grace through faith. Amen.

What Would the Sighting of Noah's Ark Mean?

Luke 16:31
"'But he said to him, "If they do not hear Moses and the prophets, neither will they be persuaded though one rise from the dead."'"

What are Christians to make of the sighting of the Ark-like object which was seen and photographed by searchers for Noah's Ark in September of 1989? We could declare, here and now, that the pictures prove the Ark has indeed been found.

But that's not how science works. We must keep our thinking clear. The accuracy of the Flood account in Scripture is not placed in doubt if this object is not the Ark. Pictures alone do not make a weighty case. An examination and detailed description of the object could make a weighty case, and there are plans to do that.

What if it is confirmed that the object on Ararat is Noah's Ark? Some Christians think that confirmation of Noah's Ark will cause millions to come to Christ. I am reminded of the Lord's words to the rich man who asked that Lazarus be sent back from the grave to warn his brothers about the judgment. Ours, too, is an evil generation which continually seeks after one more sign than it has been given.

Confirmation of the Ark, resting too high on the mountain to be explained by a local flood, would instantly make thousands of liberal theology books good for nothing more than heating fuel. Even liberal scholars would be forced to accept what they now consider to be ignorant myths of a pre-scientific age. They would be forced to deal with our message that the Word of God is true in every subject that it mentions!

Prayer: Dear Father, Your Word is true and can be trusted. Help that fact move me to spend more time with You in Your Word, reading it, studying it, and inwardly digesting it. In Jesus' Name. Amen.

Lacewing Magic

Exodus 8:19
"Then the magicians said to Pharaoh, "This is the finger of God." But Pharaoh's heart grew hard, and he did not heed them, just as the LORD had said."

The relationship between the beaded lacewing larvae and the California termite poses a real problem for those who believe there is no Creator.

The lacewing larvae enter termite-infested branches in order to finish their growth cycle. Later they leave as adults. The problem is, termites are often still at home when the larvae enter the branch. The termites, 30 times the size of the larvae, pose a serious threat to the larvae. But if you watch the larvae enter the branch, you will see them approach worker termites and wave their abdomens in the air. When they do this, the targeted termite falls right over. Upon further study, scientists have learned that the tiny larva releases a chemical while waving its abdomen, that totally paralyzes the termite. Study has also shown that the chemical works only on this termite.

If this were all a matter of chance rather than intelligent design, one wonders how there could be any lacewings left. Imagine, millions of evolutionary years ago, lacewing larvae crawling up to all sorts of creatures madly waving their abdomens at them — and then being eaten! One wonders how the larvae finally figured out their good luck; their chemical protection just happened to work on those termites which made nice safe tunnels for them to grow in.

To believe that the lacewing's defense developed just by chance is nothing more than faith. It's not science. And it is a very opposite faith from those who believe in a Creator!

Prayer: Dear Father, Your fingerprints can be seen everywhere in the creation. Give me keener sight so that I may see more of Your workmanship, and so be led to praise You more. In Jesus' Name. Amen.

Tsi-zun-hau-kau's Amazing Stick

Acts 17:26-27
"'And He has made from one blood every nation of men to dwell on all the face of the earth, and has determined their preappointed times and the boundaries of their habitation, so that they should seek the Lord, in the hope that they might grope for Him and find Him, though He is not far from each one of us."

Tsi-zun-hau-kau was a nineteenth-century Winnebago chief. A portrait of him which was painted in the last century shows him posing proudly with a stick in his hand. That stick is now at Michigan's Cranbrook Institute of Science where its carved markings have been under study.

The carvings on the four-and-a-half-foot stick are now known to be a record of the waxing and waning of the moon over a two year period. From a study of the stick, scientists have learned that Winnebago astronomers knew that the number of days in a lunar year turned out to be eleven days shorter than a solar year — and they knew this even before they ever saw the white man's calendar. To correct their lunar calendar so that it would match the solar calendar, the Winnebagos added an extra month every three years. As sophisticated as this calendar is, its structure shows that it is related to prehistoric calendars of the Mayas, Incas, and some of the ancient Siberian cultures.

Man's need to measure time, which predates other written records, shows that man is very different from the animals. The accurate and sophisticated calendars of the ancients deliver a fatal blow to those who picture ancient man as a simple, unintelligent brute who was halfway between animal and modern man. And this higher view of man fits perfectly with what the Bible tells us about man.

Prayer: Dear Lord Jesus Christ, I ask that Your people would become more skillful in showing the clear evidences that You created man above the animals. Help us to do this in such a way that will help more people seek a relationship with You. Amen.

Fish with Lungs

John 1:3
"All things were made through Him, and without Him nothing was made that was made."

The idea of fish with lungs sounds about as strange as elephants with wings. The difference is there really are fish with lungs.

In fact, there are many varieties of lungfish found in Australia, Africa, and South America. In Australia they live in quiet pools of water which become stagnant in the summer. Because there isn't much oxygen dissolved in the water, these fish need lungs in order to breathe air just as we do. Lungfish in South America live in swamps which often dry up during the dry season. When this happens, the lungfish spend the rest of the dry season resting in burrows. The African lungfish also spends the dry season burrowed in the mud. It builds a breathing tube through the mud to the surface so that he has a good supply of air.

Evolutionists have suggested that the lungfish is left over from the time when sea creatures were adapting to life on land. But their theory doesn't hold water, so to speak. Lungfish show little interest in moving to land. Besides, they lack important biological features to make that move possible. What's more, the lungfish we find in the fossil record are exactly the same as those which are living today. There is absolutely no evidence of evolution here.

While the lungfish is an unusual creature, it is a testimony to the creativity and wisdom of the Creator, and offers no support at all for evolutionary speculations.

Prayer: Dear Lord Jesus Christ, I thank You that as the Instrument of creation, You make the excellence and wisdom of God evident to man, just as You have made God's desire for a relationship with me evident through Your incarnation. Amen.

The Foam-Nesting Frog

Matthew 7:7
" "Ask, and it will be given to you; seek, and you will find; knock, and it will be opened to you."

Because there are many animals which love to eat frog tadpoles, most tadpoles never grow to be adults. If that situation was not difficult enough for the frog, some frogs live in areas where their pond is dry most of the time, preventing them from ever laying eggs.

One African frog has been given a most unique solution to these obstacles by the Creator. The foam nesting frog creates its own moist, protective nest for its young. During mating season females climb to a branch which overhangs their pond. They then secrete a mucus on the branch. All the frogs then join in kicking madly until they have worked the gooey stuff into a foam that sticks to the branch. Up to forty frogs at once have been seen doing this. Once the foam nest is complete, the frogs lay their eggs in it and return to their pond. The nest not only hides the developing young, but as it hardens on the outside, it seals moisture in with them.

The pond could dry up in a week. However, after only three or four days, because the pond could dry up in a week, the young frogs are already wiggling their way through the shell of the nest and dropping into the pond below.

The intelligent strategy of this arrangement shows us how the Creator cares for the needs of His creatures. Even if frogs had the power to evolve this amazing arrangement, what frog would be clever enough to think of such an ingenious plan?

Prayer: Dear Lord, help me to have a more clear understanding of my need to depend on You and trust that You will supply my needs. Remind me that Your wisdom and plans are immeasurably greater than my wisdom and power. Amen.

New Scientific Evidence for the Flood

Genesis 7:20
"The waters prevailed fifteen cubits upward, and the mountains were covered."

If there was a great world wide Flood, as described in Genesis, such a huge catastrophe certainly should have left obvious evidence of itself all over the world. Scientists who believe in creation say that we are surrounded by these evidences. Many of them also believe that some of the features associated with glaciers can be better explained by a catastrophic flood. While evolutionary scientists scoff at the biblical account of the Flood, they continue to run across evidences of what they take to be huge floods. Now new evidence is linking evidence for the Ice Age with a flood.

The latest example was published in the September, 1989 issue of *Geology*, a scientific professional journal. This issue of *Geology* carried a study of drumlins in northern Saskatchewan. Drumlins are hills which are traditionally associated with glacial activity. When viewed from the air, they resemble eggs lying on their sides, all pointing in the same direction.

Over the last few years some scientists have proposed that drumlins were formed from the movement of large quantities of water possibly moving beneath the ice sheets. The article in *Geology* offers yet more evidence for this theory. One of the scientists remarked that, "There's nothing in recorded history that matches the size of these floods."

If we consider Genesis to be recorded history, rather than ancient and ignorant myth, there certainly is a historical record of a huge Flood!

Prayer: In our age of unbelief, Dear Lord, so many don't even feel they need to apologize for rejecting Your Word as ancient myths. Help me, along with all of Your people, to be bright, shining lights for Your truth in this dark world. In Jesus' Name. Amen.

A Clever Golden Toad

Matthew 6:26
""Look at the birds of the air, for they neither sow nor reap nor gather into barns; yet your heavenly Father feeds them. Are you not of more value than they?"

All toads are not fat and ugly. The golden toad, which lives only in a few square miles of the cloud forest of Costa Rica, is a trim and slim toad with garishly bright, orange coloring.

The golden toad offers even more surprises. After the first heavy rain of the year, the more brightly colored males congregate around a pool of water. Often as many as 20 or 30 males gather, sitting absolutely motionless, waiting for the first female to arrive. Before long, the females begin to arrive, and less than a week later the entire population of golden toads, now numbering in the thousands, disappears back into the jungle to remain nearly invisible for another year.

This breeding strategy shows a great deal of wisdom and understanding. Have you figured it out? What the toads are doing ensures that there are plenty of golden toads in the next generation, because they have a large number of enemies in the jungle. They do this by getting together all at the same time and producing far more young than their predators can victimize in a single week.

You don't have to be a college graduate to know that toads are incapable of the planning and ability to change their physical natures as needed to carry out this strategy. Evolutionists would say that *no one* thought of this strategy. But the clear answer is that there is a Creator and He cares about all of His creatures.

Prayer: Dear Father, not a sparrow drops from the sky without Your knowledge and purpose. Not even toads lack Your care. Help me to remember Your care when I am anxious about the cares of living. In Jesus' Name. Amen.

The Mysterious Rings of Uranus

Genesis 1:16
"Then God made two great lights: the greater light to rule the day, and the lesser light to rule the night. He made the stars also."

As a result of space probes which have reached the farthest reaches of our solar system, we have learned more about the outer planets in 24 hours than astronomers learned in the past century.

One of the facts we have learned is that all five of the outer planets are surrounded by rings. Planetary rings are a wonder for another reason besides the fact that they offer the eye an otherworldly beauty. A ring comprised of dust and ice around a planet is a very unlikely and unstable structure. Every astronomer knows that the forces of the solar wind, gravity, and the unavoidable laws of physics will destroy a planetary ring within a few million years or less. Yet many astronomers believe that the solar system is billions of years old. How can this be?

To save their belief that the universe is billions of years old, astronomers have theorized that there must be small "shepherding satellites" which help hold the rings together for long periods of time. They have found satellites around Saturn which they believe may be shepherding satellites. But now an exhaustive study of Voyager 2's data from Uranus has led them to conclude that most of Uranus' rings lack shepherding satellites to hold them together. This is a real mystery to them.

Yet there is no mystery if we recognize that the Bible gives an accurate history of the entire creation, starting at the beginning, less than ten thousand years ago.

Prayer: Dear Father, in Your wisdom You have left man no place in the creation where he can look and not see Your work. But man needs more than this, he needs Your grace which only comes through faith in Jesus Christ. Help me to use Your Word to help the world know of Your Love to us in Christ. In His Name. Amen.

The Creator's Gift of Intelligence

Psalm 111:10
"The fear of the LORD is the beginning of wisdom; a good understanding have all those who do His commandments. His praise endures forever."

As if to say that intelligence is not a result of evolution but His gift to His creatures, God has given remarkable intelligence to many different animals. Some animals can perform feats of intelligence that man cannot accomplish — even if animals are limited to a much smaller area of specialty.

At the Japanese Deer Park in California, pigeons have been trained to sort electrical parts. They do this with greater efficiency and accuracy than humans. People usually quickly become bored with such work and lose their concentration. At the same park, bears have been trained to shoot baskets.

The Golden Plover migrates between Alaska and Hawaii. Yet the parents never teach their youngsters the route. When migration time comes, the young are usually old enough to fend for themselves, but too young to fly the thousands of miles, non-stop, required by the route. So the parents leave them in Alaska and head for Hawaii. A little later when they are stronger, the young plover follow the same route without error, even though they have no guide.

Intelligence is not the result of mindless natural forces, for how could mindlessness produce genius? Christians should not be surprised, as are evolutionists, to find many examples of intelligence in the animal world because intelligence is the gift of God the Creator to His creatures.

Prayer: Dear Lord, I thank You for the gift of intelligence. Help me to be a good steward of the intelligence You have given me. In Jesus' Name. Amen.

The Most Sensitive Mammal on Earth

Romans 8:28
"And we know that all things work together for good to those who love God, to those who are the called according to His purpose."

No matter where you live, whether there is a foot of snow on your yard or it is nice and warm, it is likely that a wondrous creature is, right now, at work in or near your yard. This creature has more ability to sense the things going on around him than a sophisticated space probe. It can even hear earthworms chewing beneath the ground.

This wondrous and misunderstood creature is the mole. But let's start by getting rid of some myths about the mole. Although his eyes are tiny and often hidden by fur, the mole is not, as some people think, blind. Moles don't run very well across the ground because they are made for digging. Their front feet have claws for digging which are so specialized that the mole must walk on its knuckles.

The mole's nose and tail have vibration sensors that are more sensitive than any that man's science has been able to make. These sensors are composed of thousands of parts, and allow the mole to hear and locate the grubs chewing on the roots of your lawn — even through several feet of soil! In one year's time a mole can eat almost 60 pounds of grubs — more than enough grubs to kill a fair-sized lawn.

Every one of God's creatures has a purpose, whether we understand it or not. Although you might not like the little trails which moles leave in your lawn, the trails are quite harmless. The thing to remember is that there is a mole on patrol in your yard.

Prayer: Dear Lord, I understand so little but You know and understand all things. Help me to learn more about Your wisdom and purposes in the creation, and to be strengthened when life brings things which seem to have no purpose. Amen.

The Body's Incredible Healing Powers

Mark 2:17
"When Jesus heard it, He said to them, 'Those who are well have no need of a physician, but those who are sick. I did not come to call the righteous, but sinners, to repentance.'"

God has given your body such incredible powers of healing and repair that you could almost say that each one of us has our very own doctor and pharmacy inside us.

Most people know how doctors test new medicines on people. Usually one test group receives the medicine which is being tested while another group receives a "medicine" — called a placebo — which does nothing. But doctors have known for years that something called the placebo effect is *also* taking place during these tests. For example, if the body thinks it is getting aspirin for a headache and it is actually getting a placebo, the body expects the medicine to work and the headache may often disappear. Studies over recent years have shown that the body, in fact, makes its own pain-killers. It has been found that one of these pain-killing chemicals made by the body can reduce pain as effectively as 8 milligrams of morphine or 80 milligrams of Demerol!

Studies on the body's built-in pharmacy have, so far, confirmed that the body is also able to treat coughs, anxiety, high blood pressure, depression, asthma, colds, arthritis, ulcers, high cholesterol, and even warts. Often, the medicines we take give the body "permission" to do its own healing by giving us the belief we are going to get better.

More often than not, medicine is the study of how the body solves problems — in other words, a study of the work of the Great Physician Himself!

Prayer: Dear Lord Jesus, You made us and You are the great Healer of both body and soul. I thank you for the wonderful abilities You have given my body to take care of itself. But without You, my soul has no healing. I trust in Your atonement for my sin. Amen.

A Second Set of Eyes

Exodus 4:11
"So the LORD said to him, 'Who has made man's mouth? Or who makes the mute, the deaf, the seeing, or the blind? Have not I, the LORD?'"

What has five eyes, is the only creature that stirs in the noonday sun in the Sahara Desert, and — here's a hint — is the fastest running insect on Earth? The little fellow is called the Cataglyphis, or desert ant.

The desert ant can run at a speed of up to a yard per second, or about two miles per hour. As its speed suggests, the desert ant will wander far from home — often more than 600 feet from its nest. That's a long way in ant miles. And although there are few landmarks in the desert, the desert ant never seems to get lost. Scientists have wondered how he does it.

What they have learned is that the desert ant has three eyes in addition to the usual two. These three extra eyes are three lenses in its forehead between, and a little above its normal eyes. They allow the ant to see and recognize the varying patterns of polarized light in the desert. In other words, the desert ant navigates in the featureless desert by being able to see features in light which are invisible to our eyes. One scientist observed that without its speed and the special abilities of its extra three eyes, the desert ant would never be able to survive in the desert.

It seems as if there is no end to the creative inventions and special abilities we find in the living world. That is just what we would expect, living in the creation of an unlimited, wise, and all-knowing Creator!

Prayer: Dear Father, even man's cleverest technology is poor and unimaginative when compared to the least of Your works. I thank You that You would be my perfect Father and helper through Jesus Christ. Amen.

Scientists Get Smart About Intelligence

Acts 13:6-7
"Now when they had gone through the island to Paphos, they found a certain sorcerer, a false prophet, a Jew whose name was Bar-Jesus, who was with the proconsul, Sergius Paulus, an intelligent man. This man called for Barnabas and Saul and sought to hear the word of God."

A few years ago, when we were beginning to see the first large strides in computer development, some were predicting that computers would soon have an intelligence similar to human beings. Do you ever wonder what happened to such optimistic expectations?

What happened was that scientists began to work on the problem of creating artificial intelligence. When they did, they began to learn that intelligence is not as simple as they thought it was. It's true that computers can read and speak, but they cannot understand what they are reading or speaking. They can build cars, but they cannot decide to drive to the library and increase their learning.

At the current level of development, researchers can build a computer that understands that if water continues to run in a stopped-up sink, the sink will overflow. However, by the time the computer has figured out the problem and the solution, the sink is already overflowing. We are a long way from a robot that will cook dinner for you without burning down the house.

When and if scientists ever create a computer that actually has intelligence, they will only be confirming what the Bible says about the creation. And that is that intelligence can only be the result of carefully planned and designed effort by another intelligence.

Prayer: Dear Father in heaven, I thank You for my intelligence. Fill my intelligence with Your Word so that I may be Your instrument in helping others understand that You are the Author of all intelligence, and Your Son is the Author of our salvation. In His Name. Amen.

Are All Scientists Evolutionists?

Romans 1:18-19
"For the wrath of God is revealed from heaven against all ungodliness and unrighteousness of men, who suppress the truth in unrighteousness, because what may be known of God is manifest in them, for God has shown it to them."

Evolutionists sometimes challenge those of us who believe in creation with the words, "Just name one scientist who doesn't accept evolution!" Since many Christians don't know the names of any scientists, the evolutionist then says. "See! You can't name any because there aren't any!"

At one time, most scientists were committed to the idea that there is a Creator. Then for a long time, it really did look like there were hardly any scientists who rejected evolution. Now a poll published in the February 1988 issue of the professional magazine, the *Industrial Chemist*, tells us what scientists really believe.

According to the poll of professional scientists, over one-fifth — 20.6 per cent — completely reject evolution. Less than half of the scientists — 48.3 per cent — believe that it is even possible for man to have evolved from lower forms without supernatural intervention. Do scientists think that scientific creationism is hurting science education? According to this poll, 39.9 per cent say "No."

So evolutionism is not an unchallenged "fact" of science. It seems that scientists are divided over evolution — meaning that evolution has not made its case — even to the professionals! That should help a lot of Christians understand that science does not demand their acceptance of evolution. Christians should also see that the Bible offers an intelligent alternative to evolution.

Prayer: Dear Lord Jesus, You have known, firsthand, how those who reject You will often lie to themselves and others so that they seem right and You seem wrong. We see that today as well. Give those scientists who know You the wisdom and strength to share their witness with others. Amen.

The Hand that Holds the Worlds Extended to Us!

John 1:14
"And the Word became flesh and dwelt among us, and we beheld His glory, the glory as of the only begotten of the Father, full of grace and truth."

There are some events that are so grand and awesome that we can only stand back, amazed at what we see, while comprehending that we understand so little. The birth of Jesus, the Christ, is clearly one of those events.

In the first chapter of St. John's gospel, we read that the Word which became flesh for our salvation was the Word through Whom all things were created. This Word was the Instrument that created the incredible variety of living things we see on Earth. He was the One Who created vision, the petal of a rose, the raw churning power of an atom, and the amazing brain.

It was this very Word, Who in one day, before going on to greater things, created the countless billions of stars and energized them with enough energy to shine for thousands of years. And who are we? We cannot even measure the energy those stars put out in one second of their history!

Yet on Christmas day we see the King of the universe as an infant clothed in our form, He Who makes and knows all things, resting in a manger, unable to even ask for the nourishment an infant needs or let His simplest human need be known. And why? Very simply — He was the Instrument Who made us, and we are eternally lost without His help. Who better, then, to rescue us? We can only stand in deep reverence at such love that would come among us to rescue us from our own sin.

Prayer: Dear Lord Jesus, of all that I have ever learned, the most amazing facts areYour birth, life, death and Resurrection for my salvation. Before You I know nothing and have no power; my strength and powers can only be supplied by You. Amen.

The Bear Facts About Good Health

Ephesians 2:10
"For we are His workmanship, created in Christ Jesus for good works, which God prepared beforehand that we should walk in them."

Imagine, lying around not just for days on end, but for months. And while lounging the months away, you not only don't get fatter, but you get leaner and more muscular. It sounds like a couch potato's dream.

Of course it's a good thing we can't live this way. We have been placed in this life with things to do. The fact that such laziness would cause our bones to become weak and smaller, and make the calcium level of our blood high enough to kill us, is God's way of telling us to avoid the couch potato's life style. Besides losing bone, we would also lose muscle and gain fat, lying around all day. That's why scientists were puzzled as they learned that bears don't suffer bone loss, and don't lose muscle during their winter rests. Bears don't really sleep all winter, but they do rest a lot, and bear their young. But they don't eat, drink, or go to the bathroom all winter. This combination would kill most creatures in less than a week.

Researchers believe that they have found the chemical in a bear's blood which prevents bone loss during long inactivity. The chemical is under study for possible use in treating human osteoporosis. And bears avoid going to the bathroom by chemically changing their own wastes back into proteins.

The fact that scientists study how things in the creation work to improve how we do things, is a tribute to the workmanship of the Creator, even if scientists refuse to admit His existence and activity.

Prayer: Dear Father, give strength and wisdom to those who study how You have made the creation to improve how we do things, not only for our Earthly benefit, but also that, seeing Your Hand, they would come to You through Jesus Christ. In His Name. Amen.

The Plant that Breathes Through a Snorkel

Genesis 1:11
"Then God said, 'Let the earth bring forth grass, the herb that yields seed, and the fruit tree that yields fruit according to its kind, whose seed is in itself, on the earth'; and it was so."

Some plants are designed to take the air they need in the water while others are designed to live on land. When a land plant is submerged in water, it can drown, just like one of us.

Rice is actually a land plant which needs air to breathe. Yet it must be submerged in water in order to survive, growing in water as deep as fifteen feet. In flood prone areas, rice has been known to grow as much as a foot per day in order to keep its topmost leaves out of the water. The reason that rice plants don't drown brings us to another of the Creator's dandy designs.

The rice plant is able to take advantage of physics to create a vacuum within itself, drawing in air, much like your lungs. The rice plant draws air in through its leaves, as well as through a sheath of air which surrounds its submerged stalk. Rice gives off one carbon dioxide molecule for every oxygen molecule it consumes. But because carbon dioxide dissolves more quickly in water than does oxygen, a vacuum is created within the plant, pulling in still more air. So while you could not draw air through a hose to a depth of 15 feet, the rice plant not only survives, but increases its growth rate under these conditions.

It appears that God has created so many different forms of life that there are almost no conditions on Earth where something couldn't live. That in itself is a witness that He, and not the forces of evolution, is responsible for making all living things.

Prayer: Lord, there is no scientist or anyone else who can even hope to compete with You in skill or knowledge. Help me to be a witness to this fact in a world which desires to deny that You are Creator, and thus remain ignorant of its need for forgiveness through Your Son, Jesus Christ. Amen.

Are All Dinosaurs Extinct?

Romans 8:20-21
"For the creation was subjected to futility, not willingly, but because of Him who subjected it in hope; because the creation itself also will be delivered from the bondage of corruption into the glorious liberty of the children of God."

Dinosaurs are one of the most popular subjects among grade schoolers these days. Even the U.S. Postal Service has issued a set of dinosaur stamps. These fascinating and awe-inspiring creatures seem distant and unreal to our modern world. Because of this, dinosaurs are often seen as proof of the idea that we live in an ever-changing, evolutionary world.

But are dinosaurs really extinct? Many species of dinosaurs have been discovered and classified over the years. Most of the reptiles we think of as "dinosaurs" belonged to the Archosauria or "ruling reptile" subclass. However, at least 21 species and seven subspecies of Archosauria continue to live to this day, though some of them are endangered species. The largest living group of Archosauria is the crocodiles. Most crocodiles grow to be larger than the average extinct dinosaur, which averaged about the size of a horse. Marine crocodiles found off the northern coast of Australia grow to over thirty feet in length — an accomplishment unmatched by most dinosaurs. Crocodiles help us to learn to see the extinct dinosaurs as part of God's incredible creativity, rather than as symbols of an ever-changing, ever-evolving world.

But this doesn't mean that change, especially improvement, is a myth. Our Creator wants far more for us than we can achieve for ourselves. That is why He has given us His Word of salvation in Jesus Christ! In Christ is found true and worthy improvement.

Prayer: Dear Lord Jesus Christ, my faith and trust is in You and You alone. Grant me Your instruction through Your Word that I may not only have life, but so that I may become more like You. Amen.

A New Year

Psalm 90:4
"For a thousand years in Your sight are like yesterday when it is past, and like a watch in the night."

When a year comes to a close and a new year starts, Christians have traditionally set aside time to reflect on their lives.

All of us, in his own way, usually spends at least a few moments to think about the passage of time as one year changes into another. We think about some of the events of the past year in our own lives, and we wonder what a new year will bring. Some of us will decide that last year was a better year than most; some will decide that it was a year we'd rather not repeat.

But the world does all these things, too. What, in our consideration of the passage of time, is unique about us as Christians? First of all, no matter how good or bad we think last year turned out, we know that the Lord of history is still in charge. There have been good things for which to thank Him and there have been trials. These too, He would use for our good — giving us another reason to thank Him.

When Paul preached on Mars Hill to a world very much like ours, he taught them that the Creator of all is personally involved in determining the times and boundaries of man — for a purpose. His purpose is to lead men to seek to learn more about Him, so that they might finally learn of His desire for a personal relationship with them. No matter what the new year may bring, each of us can be sure that it will include our Creator's urging to be closer to Him than we are now, for there are no bounds to His love.

Prayer: Dear Heavenly Father, I thank You for all Your blessings to me this past year. I ask that You would help me to be more aware of Your desire for a closer relationship with me through Jesus Christ from now on. Amen.